ROUTLEDGE LIBRARY EDITIONS:
WATER RESOURCES

Volume 3

THE BRITISH SEAS

THE BRITISH SEAS

An Introduction to the Oceanography and Resources of the North-West European Continental Shelf

J. HARDISTY

Routledge
Taylor & Francis Group

LONDON AND NEW YORK

First published in 1990 by Routledge

This edition first published in 2024
by Routledge
4 Park Square, Milton Park, Abingdon, Oxon OX14 4RN

and by Routledge
605 Third Avenue, New York, NY 10158

Routledge is an imprint of the Taylor & Francis Group, an informa business

British Library Cataloguing in Publication Data
A catalogue record for this book is available from the British Library

ISBN: 978-1-032-74502-2 (Set)
ISBN: 978-1-032-73599-3 (Volume 3) (hbk)
ISBN: 978-1-032-73610-5 (Volume 3) (pbk)
ISBN: 978-1-003-46508-9 (Volume 3) (ebk)

DOI: 10.4324/9781003465089

Publisher's Note
The publisher has gone to great lengths to ensure the quality of this reprint but points out that some imperfections in the original copies may be apparent.

Disclaimer
The publisher has made every effort to trace copyright holders and would welcome correspondence from those they have been unable to trace.

The British Seas

An introduction to the
oceanography and resources of the
north-west European continental
shelf

J. Hardisty

London and New York

First published 1990
by Routledge
11 New Fetter Lane, London EC4P 4EE

Simultaneously published in the USA and Canada
by Routledge
a division of Routledge, Chapman and Hall, Inc.
29 West 35th Street, New York NY 10001

Laserset by
NWL Editorial Services, Langport, Somerset TA10 9DG

Printed and bound in Great Britain by
Biddles Ltd, Guildford and King's Lynn

British Library Cataloguing in Publication Data
Hardisty, J. (Jack)
The British seas: an introduction to the new oceanography and
resources of the north-west European continental shelf.
I. Title
551.4613
ISBN 0–415–03586–4

Library of Congress Cataloging in Publication Data
Hardisty, J. (Jack), 1955–
The British seas: an introduction to the new oceanography and
resources of the north-west European continental shelf.
p. cm.
Includes bibliographical references.
ISBN 0–415–03586–4
1. Continental shelf – Great Britain. 2. Oceanography –
North Atlantic Ocean. I. Title
GC85.2.G7H37 1990   90-31652
333.91′64′0941–dc20   CIP

Contents

Plates

Figures

Figures

The computer charts are available on disk from GeoSystems Ltd, School of Earth Resources, University of Hull, HU6 7RX, UK.

Tables

Preface

This book was written to support a third year undergraduate course entitled 'Oceanography and the UK shelf', which was taken by geography and geology students at the Royal Holloway and Bedford New College, London University and is now taken by students in the School of Earth Resources at the University of Hull. The course – and hence the book – is intended to introduce the shelf region as one would a new country, exploring first its physical environment, and then summarizing its economic activity.

With practice in lecture theatres, the amount of material in each chapter has become balanced, but if there remains an emphasis it is on the detailed physical characteristics of the environment. The second half of the book then concentrates on explaining the geographical distribution of economic activity in terms of these characteristics. The book will thus serve as an introduction to any of the more specialized subjects of oceanography, marine geomorphology and geology, and maritime economics and also as an attempt to integrate the physical and economic characteristics of a new region. It is currently relevant to consider the coastal and offshore zones in this way and, since little background knowledge has been presumed, the result should be available to the widest audience.

This is not a research book, more a guide which has been compiled from the work of many other people. Field experience, if it can be called that, and which is normally an author's justification for presuming to write such a guide, was gained on fourteen research cruises in the Natural Environmental Research Council (NERC) ships RRS *Challenger*, RRS *Frederick Russell*, and RRS *John Murray* in the Celtic Sea with Dr Doug Hamilton's group from Bristol University; in the English Channel with Dr Tony Heathershaw and the Institute of Oceanographic Sciences at Taunton; with the School of Ocean Sciences, University College of North Wales in the Irish Sea; and on survey work in the North Sea. My thanks to these people, the ships, and their crews.

I am also grateful to the twelve specialists who kindly, quickly, and usefully commented on the main chapters. Their guidance through the detailed literature was important, and their own work often constitutes the main references in each chapter. Dr Colin Jago and Dr Jim Hansom read versions of the whole book, and I am grateful for their encouragement and for their advocacy of a more 'integrative' approach. I am also grateful to Kathy Roberts for typing an early draft and to Mark Daddy and Jeremy Lowe for the diagrams and computerized charts.

J. Hardisty
North Ferriby

Part I

The oceanography

Chapter one

British Seas: an overview

The human race has been a land-dwelling species since the time, some 300 million years ago, when our amphibious ancestors first evolved from air-breathing fish. However, the land masses represent much less than one-third of the surface area of planet earth, and there is an exponentially increasing requirement for space – for living, for the cultivation of crops and livestock, for the extraction of mineral and energy resources and for the disposal of the waste products from our ever more sophisticated and demanding societies. This problem is not new, and earlier cultures have dealt with increased population by migrating to expand their domains and to make fuller use of less densely inhabited areas.

It is easy to argue that a logical extension of this policy would be to colonize the 71 per cent of the earth's surface which lies beyond our coastlines, and that a book concerned with the oceanography and resources of the British Seas should provide impetus for such changes. However, the argument is fallacious because the technological investment required to inhabit such a hostile environment on a large-scale or permanent basis is presently many orders of magnitude greater than the benefits which would accrue. Instead, this book has a less pioneering but much more sensible objective and the present chapter derives and defines that objective. First, however, the boundaries of the topic must be drawn both in terms of the ocean area and of the subject matter with which this book is concerned.

Definition of the region

Although the national boundaries which cross the north-west European continental shelf will be detailed in Chapter 9, the British Seas can, for the present purposes, be defined as those continental shelf areas bounded by the coastlines of France, Belgium, Germany, Denmark, Norway, the United Kingdom, and the Republic of Ireland,

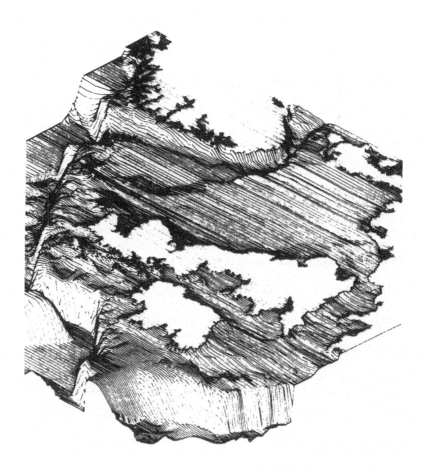

Plate 1 Computer generated chart of the north-west European continental shelf (courtesy of NERC Unit of Thematic Information Systems)

and by the continental break at a depth of 200m and between latitudes 48°N and 63°N (Fig. 1.1). Geographically, these British Seas may be divided into five regions: the North Sea, the English Channel, the Celtic Sea, the Irish Sea, and the North-Western Approaches, the last of which divides into three narrow but elongated shelf-edge regions: the Malin Sea, the Sea of the Minches, and the Sea of the Hebrides. There are thus seven seas, though one is called a channel.

Some 20,000 years ago sea-level was considerably lower than at the present time. The subsequent sea-level rise has submerged an area of more than 942,000km^2 on the continental shelf and has created some of the most important shelf seas in the world. Not only are the English Channel and the Southern North Sea, in particular, the busiest shipping clearways in the world, but the seas are also sources of the European community's industrial wealth (fisheries, petroleum, aggregates, and power) and sinks for the disposal of the refuse from its intensely urbanized and industrialized coasts.

Definition of the subject

The subject matter covered in this book divides into two discrete but interrelated areas. In Part I the oceanography of the British Seas is introduced and in Part II the resources which are exploited from this environment are detailed.

Marine scientists are called oceanographers, which derives from the Greek word *graphos* meaning 'the description of'. The modern science still carries over from its earlier days the connotation of a descriptive past. From the earliest sea voyages of the Polynesians, the requirement was for an ability to chart and to navigate vast tracts of featureless ocean. However, although early explorers occasionally recorded natural phenomena of their environment at sea, it was not until the early part of the nineteenth century that oceanography developed into a systematic science, making measurements on a routine basis and erecting models to explain the observations. The first text book on the subject was published by Matthew Fontaine Maury in 1855. It contained the first bathymetric chart of the North Atlantic Ocean, and was entitled *The Physical Geography of the Sea*. Since that time there has been a great increase in empirical and theoretical work, and we nowadays identify a number of specializations, including dynamical oceanography (concerned with the mathematical description of ocean currents), chemical and biological oceanography, marine geomorphology (concerned with the shape of the seabed), marine geology and geophysics, as well as the more applied branches of hydrographic surveying, ocean engineering, and the

5

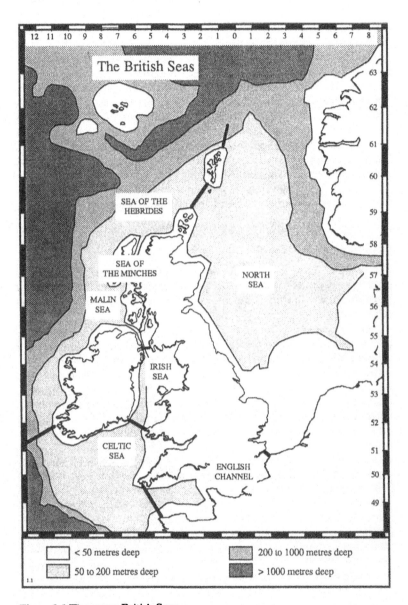

Figure 1.1 The seven British Seas.

wave and tidal specialists. Part I of this book provides an introduction to some of these fields in so far as they are concerned with the marine environment of the British Seas.

This exploration of the world ocean has developed alongside, and to a certain extent continues to service, the commercial exploitation of the ocean waters and of the seabed resources. It has been estimated that the annual income of the world's fishermen from marine catches is now in the region of £10,000 million ($15 billion) and that the world ocean-freight bill is nearly twice that figure. Again the wellhead value of oil and gas from the seabed is similar to the value of the marine fish catch, and even the relatively new seabed mining industry is currently valued at an annual £400 million ($0.6 billion). A growing proportion of this worldwide resource exploitation is conducted in the British Seas, and is detailed in Part II of this book, along with the marine power and waste disposal industries .

Definition of the object

This book is designed for the undergraduate student, or for the professional working not in oceanography or in resource management, but rather in the more general fields of environmental science, geography, geology, or marine or maritime studies. Were such a reader to wish to learn about some town, state, country or even continent it is likely that an enquiry at a local bookshop or library would produce any number of suitable publications. These would detail, on the one hand, the rocks, relief, climate, and botany of the region, and on the other hand the distribution of population, industry, transport, and commerce. Despite the fact that the world ocean occupies the majority of the earth's surface, such an enquiry would not be successful if details of an offshore region were sought. An initial objective of this book was, therefore, to fulfil that role for a specific sea area, that is the British Seas. However, it became apparent that the product of such an objective is little more than a word atlas or gazeteer, made dull and lifeless by factual repetition. Better then, if some of the oceanology (from the Greek word *logos* which means 'the logic of' or 'the science of') of the environment can be explained so that the British Seas can be understood rather than simply described.

The objective of the first half of this book is therefore to provide a sound and relevant introduction to the sciences of marine geomorphology, marine geology, and oceanography and to exemplify these considerations with the waters of the British Seas. The chapters are designed so that the reader will be presented with sufficient information to allow a detailed description of the large-scale features to be comprehended, together with guidelines for further reading in accordance with individual taste.

The second half of the book then draws heavily upon this environmental background for it attempts to answer one simple question: 'Why is this resource exploited at that location in the British Seas?'. Conceptually there are very many methods which may be chosen to answer this question. This book, however, utilizes what will be termed a 'spatially restrictive' approach. For each of the six resources which are considered in Part II, the preliminary assumption is that exploitation could be accomplished in each one of the 942,000km² of the British Seas. There are, however, three sets of restrictions which must be considered, and which limit the distribution of resource exploitation.

First, consideration of one or more of the environmental parameters which are detailed in Part I will immediately eliminate certain areas. Gravel cannot be extracted from a solid rock floor which is swept clean by modern tidal currents, for example, nor can much tidal power be generated from the small tidal ranges of the southern North Sea. These will be termed the 'environmental restrictions', and they clearly limit the distribution of successful resource exploitation.

Second, the extraction, processing, and marketing of the resource each require favourable financial returns before investment and operation can continue. These will be termed the 'economic restrictions' of the resource.

Finally, although the resource may be present in abundance at a particular location, and although the economics may be favourable, there are frequently legislative restrictions placed upon resource exploitation which may involve national or international issues and may be invoked because of wider scale environmental or economic issues. These will be termed the 'policy restrictions' of the resource.

This analysis shows that certain resources are already over-exploited, whilst others are perhaps under-utilized and these general conclusions are drawn together in the final chapter.

Structure of the book

The book is divided into two sections, and each section is divided into seven chapters. The first and last chapters are brief and are exceptional for they deal separately with the conceptual analysis of the British Seas. The present chapter has explained how the spatially restrictive analysis works, and the last chapter summarizes the results of that analysis through tables of conclusions which can be found on pp. 246–9. The remaining twelve chapters are equally divided between the oceanography and the resources of the region.

Part I, which deals with the oceanography of the region, commences with an orientation chapter ('The shape of the shelf') which is designed to provide the necessary signposts for the sea areas and seabed features which will be encountered on the following pages. The evolution of the region is detailed in Chapter 3, 'Geological history', which not only provides a literal foundation for the oceanography but also explains the environmental potential for the hydrocarbon industry which will be covered in Chapter 8. Chapters 4 and 5, 'The wave regime' and 'The tidal regime', deal with the two main ocean energy inputs into the British Seas system, and provide the necessary background for the seabed sediment work covered in Chapter 7, and for the marine power industry covered in Chapter 12. These are followed by Chapter 6, 'The oceanographic regime', which deals first with the distribution of temperatures and second with the distribution of salts in the British Seas, and specifically with annual variations in those distributions. This provides the necessary background information for an analysis of the distribution of fish species in Chapter 10. An introduction to biological oceanography has been deferred until Chapter 10, and has been restricted to fish biology for that reason. The last chapter in Part I, 'Modern shelf sediments', deals with the distribution and movement of sands and gravels in the British Seas. It has been placed at the end of Part I because it utilizes tidal theory to explain the seabed sand transport paths, and is itself referenced by the mining industry chapter later in the book. Each of these oceanographic chapters has a similar structure, in that a description of the techniques which are employed to measure environmental parameters is followed by a summary of the results for the region as a whole and concluded with an explanation of the observed distributions in terms of the oceanographic processes.

Part II, which deals with the resources of the region commences with 'Trade and shipping', in which the main shipping routes are formally related to port locations and cargo handling specializations and to the navigational requirements of the British Seas. This is followed by 'Hydrocarbons' in which the development of the resource is related to the submarine geology detailed in Chapter 3, and then to economic and legislative controls. Chapters 10 and 11, 'Fishing' and 'Seabed mining', follow a similar pattern and refer back to Chapters 6 and 7. Chapter 12, 'Wave and tidal power', utilizes the description of the distribution of wave energy given in Chapter 4 and of tidal energy given in Chapter 5. The last full chapter, 'Waste disposal', treats the problem of pollution in terms of resource utilization and builds on the description of the net water circulation patterns in the British Seas given in Chapter 6. Each of the resource chapters also follows a similar pattern, although this is of course different from those in Part I. Each begins by examining the industrial technology and the environmental potential of the resource and then analyses the economic and legislative restrictions to attempt to explain the distribution of exploitation across the shelf.

It is apparent, then, that, material covered in this book is of a cumulative nature, with many instances of later work requiring an understanding of earlier chapters, so that the overall question, 'Why is this resource exploited at that location in the British Seas' can be properly addressed.

Additional material

The oceanographic and resource literature which deals with the British Seas is already extensive and is expanding at a seemingly exponential rate on a daily basis. Each of the following chapters is based upon a small number of review texts and these are referenced in the appropriate chapter introductions. These texts form a second tier of reading and are recommended to the interested student. A third tier consists of individual research papers which are again referenced in the text and should be addressed only for very specific details on certain data or research results. However, there is one additional source which should be at the reader's elbow throughout, and which has proven to be so relevant and useful that it has become the practice to provide each student of the course on which this book is based with individual copies. The publication is the *Atlas of the Seas around the British Isles* (ISBN 0 907545 00 9), which is produced and revised by the Ministry of Agriculture, Fisheries and Food. The *Atlas* is available from the MAFF Atlas Office, Fisheries Laboratory, Pakefield Road, Lowestoft, Suffolk, NR33 0HT, UK. The telephone number of the laboratory is (from within the UK) 0502 62244.

Chapter two

The shape of the shelf

In this chapter, we define more carefully the boundaries and outline the present-day geomorphology of the different regions of the British Seas in order to provide signposts for the sea areas and seabed features that will be encountered on later pages.

The shape of the submarine landscape is known as the seabed bathymetry, and bathymetric surveys have been conducted in the British Seas for at least the last two centuries. Bathymetric surveying demands two practical requirements, both of which have, until relatively recently, been difficult to achieve. The surveyor must first locate his position on the surface of the sea and second determine the water depth beneath the sea surface so that contoured charts can be drawn up which describe the seabed bathymetry. The principles of position fixing and depth recording will be explained before progressing to the detailed bathymetries of the five regions of the shelf.

Position fixing at sea

There is a wide variety of methods for offshore position fixing which vary from the simple optical sextant fix to electronic systems linked to orbiting satellites. The following description is based on Hooper (1979). The methods of position fixing may be divided into six groups:

1. Optical systems
2. Hyperbolic systems
3. Microwave systems
4. Range and bearing electronic systems
5. Combined electronic and optical systems
6. Satellite systems

Close to the coast the observer may use simple compass bearings or a vertical sextant to determine the distance from a coastal land-

The oceanography

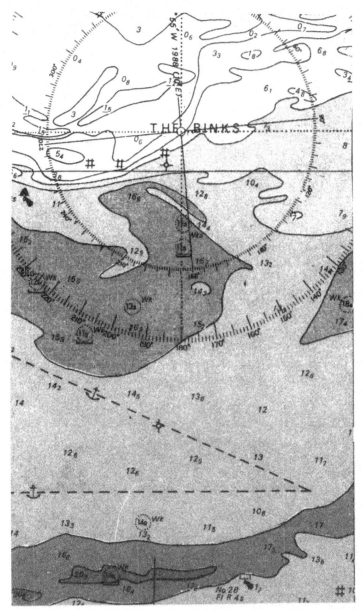

Plate 2 Hydrographic charts provide a wealth of information on seabed
bathymetry and sediments (courtesy of the hydrographer of the Royal Navy)

12

mark. With less accuracy, celestial observations with a sextant can be combined with careful timing and tables of astronomical constants to determine the ship's position at sea. These optical systems demand good visibility and a relatively calm sea; only in the hands of an experienced seaman will reliable results be obtained from a heaving deck.

The hyperbolic systems are based upon the comparison of signals from shore-based radio stations. The Decca Navigator system is the primary aid of this type for ships operating in the seas around the British Isles. It consists of chains of radio stations which transmit low-frequency radio signals. The signals intersect in a known hyperbolic pattern (Fig. 2.1(a)) and a receiver on the ship displays the results on three Decometers as sets of letters and numbers. Each letter and number identifies a line on the chart and the intersection of lines from two of the Decometer readings allows a position to be plotted.

The microwave, range and bearing, and combined systems are most useful in coastal locations and effectively replace the line of site optical systems by providing electronic distance and direction information with respect to shore-based transmitters. Most recently, satellite navigation systems have been introduced which measure the distance and direction of one or more orbiting or geostationary satellites. The systems also gather information on the height and position of the satellites, and can then compute the ship's position with remarkable accuracy. It is likely that satellite navigation systems will supersede all of the other position-fixing methods in the future.

Depth recording at sea

The depth of water at any given location has, until recently, been determined by the traditional lead and line soundings. In shallow water this proved sufficiently accurate for most purposes and tallow wax was often inserted into the base of the lead to recover a sample of seabed sediment (Chapter 7). However, the technique is laborious and can, in deep water, be extremely time consuming and prone to inaccuracies. Linklater (1972), writing about the famous voyage of the *Challenger* in 1872, notes that a single deep sea depth measurement often consumed a whole day's work and that the errors due to a non-vertical line or wire could be considerable.

The advent of underwater acoustics in the early twentieth century led to the development of echo-sounders which measure the two-way travel time of a pulse of sound from the vessel to the seabed and back, and convert this to water depth. Hydrographic echo-sounders usually

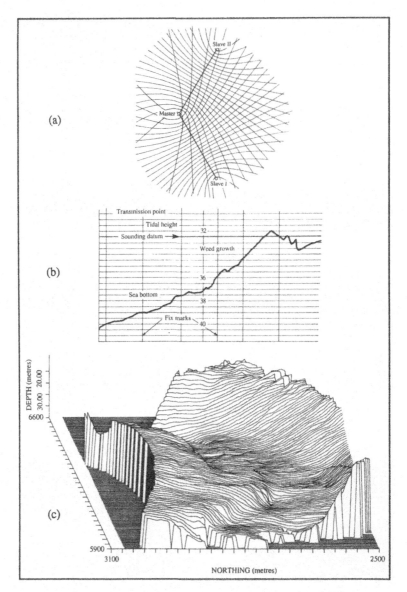

Figure 2.1 Techniques of position fixing and depth recording: (a) Decca hyperbolic chart system, (b) an example of the output from an echo sounder, and (c) bathymetric swathe survey produced by the Bathyscan system.

14

employ a belt-driven or radial-arm-driven electrical stylus passing across a strip of specially prepared paper to provide a continuous profile of the seabed as shown in Fig. 2.1(b). They employ a higher stylus speed, and hence faster pulse repetition rate, than conventional navigational echo-sounders, in order to provide maximum resolution of the seabed and a comparatively large vertical scale. Typically, transmission frequencies lie in the band 30–210kHz. Lower frequencies are used in deep water because of the attenuation of the higher frequencies.

The demands on hydrographic surveying for navigation purposes are being exacerbated by the large drafts of modern bulk carriers (Chapter 8). Existing ships of 400,000 tons draw some 27m of water, and there are proposals for ships of 800,000 tons which will have a draft of more than 30m. The whole of the shelf must be regarded simply as harbour approaches for such vessels, since large parts of the North Sea, for example, are less than 31m deep. The problem is made more complex by the presence of mobile or semi-mobile sand waves (Chapter 7) which are often many metres shallower than the surrounding seafloor. Their crests can lie at wavelengths of more than 150m, and consequently they can be missed by surveys which happen to run parallel to the crest lines. A very modern solution to the problem lies in the development of three-dimensional echo-sounders which automatically produce contoured charts of the seabed over quite wide swathes (Fig. 2.1(c)), rather than the simple depth beneath the ship which is generated by conventional echo-sounders. The associated technique of side-scan sonar is described in Chapter 7.

Physiographic features of the shelf

These types of position-fixing and depth-recording techniques have, over the past half century, resulted in the construction of the first reasonably accurate charts of the region. The physiographic features in each of the five main areas are described in the following sections.

The North Sea

The North Sea is an epicontinental sea lying on the continental shelf of north-west Europe. It is virtually surrounded on three sides by land (Figs 2.2 and 2.3) being open to the Atlantic Ocean only to the north, where it continues into the Norwegian Basin, and to the Baltic Sea in the east, and the English Channel to the south through the narrow but important Strait of Dover.

During the greater part of the last 2 million years, known as the

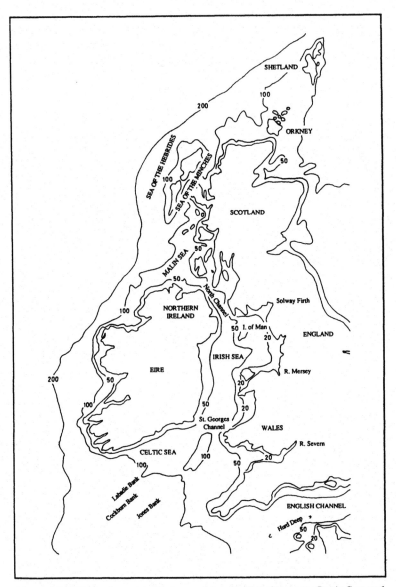

Figure 2.2 Main bathymetric features of the English Channel, Celtic Sea and Irish Sea.

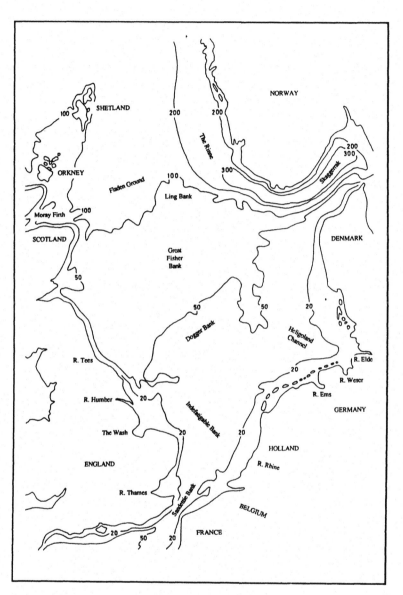

Figure 2.3 Main bathymetric features of the North Sea.
Source: Caston (1979).

Quaternary (Chapter 3), the area was directly affected by glaciation, and deposits from successive glacial advances are well-documented from bordering countries. The following bathymetric introduction to the region is based on Caston (1979). The southern North Sea is understandably the better known, with a history of surveying which dates back to the early nautical charts constructed by sixteenth and seventeenth century Dutch cartographers. Large areas to the south of 56°N have been accurately surveyed by echo-sounder since the 1930s, but the great majority of depths shown on navigational charts north of this latitude are based upon isolated lead-line soundings from the eighteenth and nineteenth centuries. These earlier surveys suffered from inadequate navigation techniques, and the errors in depth determination discussed above, so that the true distribution and shape of many submarine features, especially in the northern North Sea, is imperfectly known. Results of all significant surveys to the mid twentieth century were summarized by Stocks (1955) for the area south of 55°20'N and by Flinn (1973) for the area between 56°N and 63°N. Figure 2.3 is based on Caston (1974, 1979) and covers the region of interest here. The map shows the major relief features, and these will be discussed in terms of four principal areas:

Shelf edge

The 200m isobath which corresponds approximately to the shelf-break off the west coast of Scotland, trends in a north-easterly direction past the Shetland Isles as far as 61°37'N. Here it turns to run approximately west–east and deepens towards a fairly subdued break in slope at a depth of nearly 300m on the western margin of the Norwegian Channel. The shelf edge crosses the mouth of the Norwegian Channel at a depth of some 450m, and then swings gradually round to run in a north-north-easterly direction up the west coast of Norway.

Norwegian Channel

This deep submarine valley, which is also known as the Rinne, is the largest single feature on the continental shelf and runs for some 700km down the west coast of Norway before curving eastwards and ending abruptly in the Skagerrak. It is, in plan view, an elongate funnel diminishing in width from some 180km at the shelf edge in the north, to about 45km off the Skagerrak. It is relatively straight-sided and flat-floored with a central depth typically between 250 and 300m, but deepening at its northern end to some 450m, and plunging to over 700m off the south coast of Norway.

Northern North Sea

This area, lying between the British coast, the Norwegian Channel, and from 56°N northwards to the edge of the continental shelf, deepens rapidly to about 100m away from the Scottish coast and islands, and the greater part of the area lies between depths of 100 and 140m. The seabed is, however, far from flat and evidences a pattern of shallower banks with depths from 60m in the south to 100m in the north, separated by deeper areas of between 120 and 160m.

Despite the irregular nature of these features, some generalizations can be made. In the first place, the average depth increases significantly by about 35m to the north of a line between the north-eastern extremity of mainland Scotland, and the south-western extremity of Norway. Second, depths are greatest, at about 145m (Fig. 2.3), along a central zone lying roughly between 0° and 1°30'E, the seabed on either side rising both towards the coast and towards the western margin of the Norwegian Channel; this margin lies at a depth of less than 120m as far as 60°N. Finally, the region north of about 61°20' does not show the irregularities and undulations which so characterize the bed farther south. Additionally, there are narrow deeps or holes (Swallow, Devils, Swatchway) with maximum depths in excess of 200m found principally within an arcuate belt with its apex to the north of Aberdeen; these have been attributed to sub-glacial erosion (Flinn, 1967). Later geophysical studies have shown that infilled linear deeps are abundant in the Quaternary sediments throughout much of the central and northern North Sea.

Southern North Sea

The southern North Sea, the German Bight, is bounded to the east and the south by the Danish and German coasts, to the north by the line of banks at about 55°30'N, and extends eastwards as far as the Dogger Bank at approximately 3°E. It is a relatively shallow and featureless area ranging from 30 to 50m in water depth. The coastal waters are extremely shallow, and the seabed deepens only gradually to the north and away from the shorelines. A notable feature of the area is the abundance of sandbanks which extend as a series of parallel, linear features in 30–40m water depth up to 100km offshore. These features are described in detail in Chapter 7.

The English Channel

The seafloor to the south and west of Britain can be divided into the English Channel and the Celtic Sea respectively (Fig. 2.2). The outer, western English Channel and the southern Celtic Sea are known as the South-Western Approaches. The continental shelf of the South-Western Approaches is one of the widest in north-western Europe and extends some 900km eastwards along the English Channel to the Strait of Dover. The following description is based upon Hamilton (1979).

The Channel is bounded by the Strait of Dover in the east, the French and English shorelines and by a rough line between Cornwall in the north-west and Brittany to the south-east. The axial regional gradient is extremely low (1:5,500) whilst transverse sections show similarly low gradients except close to the present coast or where localized depressions occur. The seabed is characterized mainly by outcrops of soft, easily eroded Mesozoic, Tertiary, and Quaternary sedimentary rocks (Chapter 3) and by mobile sediments largely derived from them, or from biological production (Chapter 7). The same is largely true of the shelf seabed in most of the other areas to the north except that they possess extensive covers of glacial and interglacial deposits. The Channel is shallow and deepens gently from about 30m water depth in the Strait to about 70m in the west. Close to the shore, the seabed profile along the whole Channel may be cliffed or show stepwise changes in gradient. These are related to coastal erosion during periods of lower sea-level during the Late Tertiary and especially during the Pleistocene glaciations (Cooper, 1948; Wood, 1974, 1976; Donovan and Stride, 1975 and Chapter 3).

The Channel, however, shows evidence of some of the most significant recent erosional features on the shelf's surface. A number of deeps occur, the largest of which is the Hurd Deep. This feature is some 200km in length, more than 5km across and is from 50 to 80m deeper than the adjacent seafloor. Its shape is rather sinuous and it lies roughly along a line to the south-west from the Strait of Dover.

Sub-bottom acoustic profiling (e.g. Hamilton and Smith, 1972) has revealed that these depressions are the partly infilled remnants of much more deeply eroded features and that other connected features appear to be associated with the Hurd system. Palaeovalleys have now been found from the Dover Strait and River Seine. These have been infilled, but they once joined the main partially buried channel north of the Contentin Peninsula. Hamilton and Smith (1972) demonstrated that these formed a large, integrated fluvial system receiving drainage from much of north-west Europe during the maximum withdrawal of the sea in the Quaternary.

The Celtic Sea

The Celtic Sea (Fig. 2.2) is here defined as the body of continental shelf water lying south of a line from St David's Head to Carnose

Point, and to the west of Lundy Island in the Bristol Channel, and of a line between Land's End and Ushant in the English Channel. It covers some 75,000km².

Four distinct morphological regions can be identified along a transect taken from the north-east to the south-west of the region. The north-eastern sill (Belderson and Stride, 1966) composes a relatively flat, featureless seabed with depths between 90 and 100m, which deepens towards St George's Channel to the north. There is some evidence of submerged cliffs near the Cornubian coast (Wood, 1976) and the Haig Fras granite outcrops in broad irregular elevations which rise to about 20m above the surrounding area (Evans and Fletcher, 1987). The low-relief topography is evidently due to marine planation, completed during the Quaternary (see Pantin and Evans, 1984 and Chapter 3 for additional information). Repeated marine erosion and deposition during changes of sea-level, has evidently obliterated any subaerial topography which developed during low sea-level.

Further south-west the bed comprises the elongate, parallel, north-easterly trending ridges and troughs of Great Sole, Cockburn, Jones, and Labadie Banks whose crests lie between 90 and 100m below the surface with interbank troughs some 55m deeper. These are known as tidal sand ridges (TSRs) and are attributed to tidal controls at a time of lower sea-level (Huthnance, 1982; Pantin and Evans, 1984).

South of approximately latitude 50°N the ridge and trough topography becomes more irregular, and the banks less elongate with depths to about 160m. The western boundary of the Celtic Sea is marked by the continental shelf edge and slope, dissected by a series of steep walled canyons (King Arthur, Whittard, Shamrock, and Black Mud Canyons) where the depth increases from about 160m to the 5,000m of the abyssal plain.

In the northern Celtic Sea, there is a large, shallow depression, elongated north-north-east–south-south-west and measuring approximately 120 by 30km, known as the Celtic Deep. It descends to a depth of 20–40m below the general level of the seabed and appears to be due to continued tectonic subsidence of an older basin (Chapter 3).

The Irish Sea

The general bathymetry of the Irish Sea (Fig. 2.3) is described by Bowden (1979). This region is taken to comprise the sea area extending from a line joining Carnose Point and St David's Head in the south, to the North Channel between Larne and Corsewall Point. It thus includes St George's Channel (75–140km wide and 150km long) between Ireland and Wales and the broader northern area (195km from east to west and 150km from north to south) with the Isle of Man in the centre. At the southern end, St George's Channel communicates with the Celtic Sea which is itself open to the Atlantic, whilst at the northern end communication is again maintained with the deep ocean through North Channel, although this is, at its narrowest, only 20km wide between Tor Point in County Antrim and the Mull of Kintyre. The sea has a total area of about 47,000km^2.

Much of the Irish Sea is comparatively shallow, although a channel deeper than 80m can be traced northwards through St George's Channel and west of the Isle of Man to the North Channel. It is probable that these features can be attributed to fluvio-glacial erosion at a time of lower sea-level. The north-eastern part of the Sea is mostly less than 55m deep and between the Isle of Man and the English coast the mean depth is only 30m compared with a depth of 130m between the west side of the Isle of Man and Ireland. The deepest part of the Irish Sea is found in the North Channel west of Galloway where it reaches nearly 275m. Off the north-west coast of England there are large areas of sandbanks which are exposed at low tide, particularly in the Solway Firth, Morecambe Bay, and the estuaries of the Ribble, Mersey, and Dee.

The North-Western Approaches

The North-Western Approaches (Fig. 2.3) form the narrow but deep continental shelf region between the mainland coastlines of Northern Ireland and Scotland and the 200m isobath. The region is divided into three seas. The southernmost is the Malin Sea lying north of Donegal and Northern Ireland and south of approximately 57°N. North of the Malin Sea, the region is divided by the Outer Hebridean islands into the Sea of the Minches which lies between the islands and mainland Scotland and the Sea of the Hebrides which lies between the islands and the shelf break.

The morphology of the Malin Sea varies considerably in the area west of Islay and Mull. West and north-west of Islay, the seabed descends from about 40m to some 140m in a series of broad undulations with a wavelength of from 10 to 30km and an amplitude of 80m. Here the Stanton Banks and Blackstone Bank correspond to resistant older rocks. Further south, a south-east-facing scarp corresponds to the Skerryvore Fault (Evans *et al.*, 1986) and increases from a height of about 40m south of Tiree to some 180m south of Coll. Immediately west of Mull, there occurs another strong scarp slope which corresponds to the boundary between Tertiary basalts, and the nearby Permian to Cretaceous formations (for further information see Evans *et al.*, 1986 and the details of the geological history of the shelf given in Chapter 3).

The Sea of the Minches consists of two large interconnected channels, with adjoining lochs and sounds, between the mainland of Scotland and the Hebridean Islands. Being close to the mountains that once nourished Britain's largest and most powerful glaciers, this part of the Scottish continental shelf has suffered extensive erosion and is now covered in many places with glacial deposits. The latest bathymetric charts of the Minches (Bishop and Jones, 1979) show that the most important influence on the bathymetry has been a well-developed pre-glacial drainage system. These fluvial features on the seafloor were first mapped by Ting (1937) and in greater detail by Goddard (1965). The rivers of the Northern Highlands today are but the short and fast flowing upper courses of the drowned mature sections.

In the Sea of the Hebrides the bottom takes the form of a broad, asymmetric channel running approximately north-north-east–south-south-west and descending to a maximum depth of 240m in the west. The channel is subdivided by several banks with deep local depressions, and the banks appear to be due to resistant Tertiary bedrock. The western edge of the channel is a well-defined scarp which corresponds to the Minch Fault (Evans *et al.*, 1986) and which reaches a height of 140–160m south of Barra.

References

Belderson, R.H. and A.H. Stride, 1966. Tidal current fashioning of a basal bed. *Mar. Geol.*, 4, 237–257.

Bishop, P. and E.J.W. Jones, 1979. Patterns of glacial and post-glacial sedimentation in the Minches, North-West Scotland. In: Banner, F.T., Collins, M.B., and Massie, K.S. (Eds) *The North West European Shelf Seas: the Sea Bed and the Sea in Motion*, Vol. I, *Geology and Sedimentology*. Elsevier, Amsterdam, pp. 89–194.

Bowden, K.F., 1979. Physical and dynamical oceanography of the Irish Sea. In: Banner, F.T., Collins, M.B. and Massie, K.S. (Eds) *The North West European Shelf Seas: the Sea Bed and the Sea in Motion*, Vol. II, *Physical and Chemical Oceanography and Physical Resources*. Elsevier, Amsterdam, pp. 391–465.

Caston,V.N.D., 1974. Bathymetry of the Northern North Sea. *Offshore*, 34, 76–84.

Caston,V.N.D., 1979. The Quaternary sediments of the North Sea. In: Banner, F.T., Collins, M.B., and Massie, K.S. (Eds) *The North West European Shelf Seas: the Sea Bed and the Sea in Motion*, Vol. I, *Geology and Sedimentology*. Elsevier, Amsterdam, pp. 195–270.

Cooper, L.H.N., 1948. A submerged ancient cliff near Plymouth. *Nature*, 161, 280.

Donovan, E.T. and A.H. Stride, 1975. Three drowned coastlines of probable late Tertiary age around Devon and Cornwall. *Mar. Geol.*, M35–M40.

Evans, C.D.R. and B.N. Fletcher, 1987. Haig Fras including part of Labadie Bank. Sheet 50°N 08°W. *British Geological Survey*. 1:250,000 Series. Solid Geology.

Evans, D., R.J. Whittington and M.R. Dobson, 1986. Tiree. Sheet 56°N 08°W. *British Geological Survey*. 1:250,000 series. Solid Geology.

Flinn, D., 1967. Ice front in the North Sea. *Nature*, 215, 1151–1154.

Flinn, D., 1973. The topography of the sea floor around Orkney and Shetland, and in the northern North Sea. *J. Geol. Soc. Lon.*, 129, 39–59.

Goddard, A., 1965. Recherches de geomorphologie en Ecosse du Nord-Ouest. *Société d'Editions de Belles Lettres*, Paris. 701 pp.

Hamilton, D. 1979. The geology of the English Channel, South Celtic Sea and Continental Margin, South Western Approaches. In: Banner, F.T., Collins, M.B. and Massie, K.S. (Eds) *The North West European Shelf Seas: The Sea Bed and the Sea in Motion*, Vol. I. Elsevier, Amsterdam, pp. 61–87.

Hamilton, D. and A.J. Smith, 1972. The origin and sedimentary history of the Hurd Deep, English Channel, with additional notes on other deeps in the western English Channel. *Mem. Bur. Rech. Geol. Min.*, 79, 59–78.

Hooper, D.J., 1979. Hydrographic surveying. In Dyer, K.R. (Ed.) *Estuarine Hydrography and Sedimentation*, Cambridge University Press, Cambridge, pp. 41–56.

Huthnance, J.M., 1982. On one mechanism forming linear sand banks. *Est. Coast. Shelf Sci.*, 14, 79–99.

Linklater, E., 1972. *The Voyage of the Challenger*. Cardinal, London, 288 pp.

Pantin, H.M. and C.D.R. Evans, 1984. The Quaternary history of the central and southwestern Celtic Sea. *Mar. Geol.*, 57, 259–293.

Stocks, T., 1955. Der Boden der sudlichen Nordsee. 2. Beitrag: Eine neue Tiefenkarte der sudlichen Nordsee. *Dtsch. Hydrogr. Z.*, 9, 265–280.

Ting, S., 1937. The coastal configuration of western Scotland. *Geog. Annal.* 19, 62–83.

Wood, A., 1976. Successive regressions and transgressions in the Neogene. *Mar.Geol.*, 22, M23-M29.

Chapter three

Geological history

During the last few decades, the geological understanding of north-west Europe and its vast continental shelves has advanced tremendously as a consequence of research efforts and surveys carried out by academic institutions, national geological surveys, and government-sponsored oceanographic institutes, but most importantly due to the exploration activities of the oil industry. Intensified study of the classical outcrop areas has led to the development of new stratigraphic and structural concepts. This has been married with marine geophysical surveys supported by deep-sea drilling, and has resulted in increased knowledge of the geology of the oceans and continental slopes flanking mainland Europe. This wealth of new data, in combination with the geology of the classical outcrop areas and the oceans now permits us to reconstruct the geological evolution of north-west Europe in a modern, plate tectonic framework. This story is best understood in geochronological terms by reconstructing the palaeogeography of the region at a number of important stages in the past 500 million years.

The present chapter examines the development of these new ideas in order to provide the literal foundation for the oceanographic work which is to follow. Additionally, this chapter describes the origin of the hydrocarbon reserves of oil and gas in the British Seas to provide the necessary background for Chapter 9, 'Hydrocarbons'. Finally, the chapter concludes with a description of the region during the last 2 million years, and this forms the basis for the analysis of modern shelf sediments in Chapter 7 and of the offshore mining industries in Chapter 11.

Plate tectonics and geological time

Perhaps the most important development in the geological sciences this century has been the acceptance of the idea that the continents

Plate 3 The geological history of the region is contained within the rock record which is exposed on land or from offshore survey work (courtesy of Michael Jay Publications)

of the earth were once joined in a massive super-continent called Pangea (Fig. 3.1). The last 300 million years of earth history, and the geological strata which corresponds to this time, can then be explained by the idea that Pangea has broken into a number of smaller continents which have been forced across the surface of the planet by thermal currents deep within the earth. These currents rise to the surface along mid-ocean ridges, and new oceanic plates are formed which push sideways and widen the ocean basins. The mid-ocean ridges are called constructive plate margins, and the Mid-Atlantic Ridge system stretching from Iceland to the Antarctic Ocean is one modern-day example. Conversely, at the edges of continents, oceanic plates are forced downwards beneath the lighter continental plates, developing deep submarine trenches, and resulting in flexures and foldings of the continent. These zones of collision are called destructive margins; the Marianas Trench in the western Pacific is again a modern example.

The geological history of the British Seas is best understood in terms of these global changes; geologists divide the vast intervals of time which are involved into four distinct 'eras'. The oldest era is called the Pre-Cambrian, and represents the time from the formation of the planet some 5,000 million years ago up until the earliest forms of life some 570 million years ago. This is followed by the Palaeozoic era, from the Greek words *palaios* meaning old and *zoe* meaning life, which lasted until about 225 million years ago, and was in turn followed by the Mesozoic era (from the Greek *mesos* meaning middle). Finally, and most recently, is the Cenozoic era (or sometimes the Cainozoic, from the Greek *kainos* meaning new) which began about 63 million years ago and continues to the present day. The long eras of geological time are in turn divided into shorter intervals known as periods, and these are listed in Table 3.1. The geological history of the British Seas includes events which date back into the Pre-Cambrian, however, for the present purposes, it can most usefully be traced from the Devonian, Carboniferous, and Permian periods in the late Palaeozoic. The break up of Pangea in the early Mesozoic (Fig. 3.1) began with a major fracture zone which developed between what is now the Americas and Africa and eventually opened to form the Atlantic Ocean. Other cracks opened at approximately the same time between East Africa, India, Australia, and Antarctica, producing the Indian Ocean. The Atlantic opened gradually to its present width and, in the Cenozoic, Greenland separated from North America forming Baffin Bay.

The environment in which the rocks of the British Seas were laid down also changed during these global movements. In Permian times the region lay close to the equator and the continental desert sands

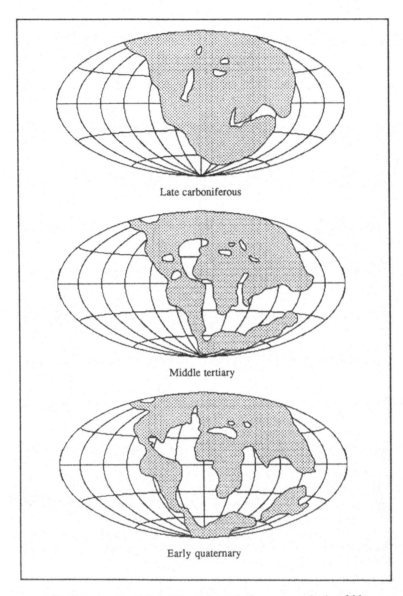

Figure 3.1 Development of the North Atlantic Ocean over the last 300 million years.

28

Table 3.1 The periods of geological time

Date (million years BP)	Period	Movements in Britain (and in the Atlantic)
2	Quaternary	Divides into the Pleistocene and Holocene epochs. Ongoing E–W tilting of England. (*Atlantic continuing to spread outwards from central ridge.*)
26	Neogene	Divides into the older Miocene and younger Pliocene epochs. Flexuring and slight elevation of England as a result of Alpine movements.
63	Palaeogene	Divides into the oldest, the Palaeocene, the Eocene and the youngest, the Oligocene epochs. General worldwide elevation. (*Separation of Europe and Greenland and shelf boundary defined.*)
136	Cretaceous	Began with a land epoch in Britain followed by the Cenomanian transgression, and finished with a period of maximum extent of the seas and stability in Britain. (*South America began to separate from South Africa (c. 100 million years)*).
195	Jurassic	Began with downwarping in Britain followed by general elevation. (*North America and Africa began to separate, followed by initiation of North Atlantic (c. 180 million years)*).
225	Triassic	Began with continued erosion in Britain, and basin infilling followed by the Rhaetic marine transgression.
280	Permian	Began with the uplift of the Pennine block, followed by rapid denudation and faulting and a general lowering of sea-level.
345	Carboniferous	Began with a marine transgression into Britain from the south, followed by complex elevation.
410	Devonian	Culmination of folding and warping episodes followed by local sedimentations.
440	Silurian	Final infilling of the Caledonian basins
530	Ordovician	Began with intense folding and metamorphism in the north, followed by downwarping of ocean floor and finally elevation and local folding in Britain.
570	Cambrian	Began with a marine transgression followed by seafloor spreading and sedimentation, and ended with some elevation in Britain.
4600	Pre-Cambrian	Lewisian and Torridonian rocks of north-west Scotland date from Pre-Cambrian times.

which were deposited at that time are analogous to the arid regions of the present-day Sahara and Sahel deserts. The region has since progressed steadily northwards, lying in tropical latitudes during the Triassic and Jurassic, and the chalk of the Cretaceous was deposited in a sub-tropical sea some 60 million years ago. The following sections are based on Ziegler (1982) and reconstruct the

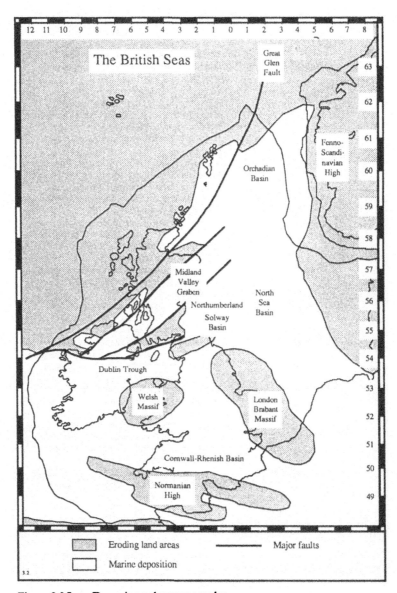

Figure 3.2 Late Devonian palaeogeography.

palaeogeography of the British Seas region throughout these changes.

Devonian palaeogeography

A drilling rig working at the present time in the British Seas is likely to encounter rocks from any or many of the geological periods which are listed in Table 3.1. In general, this is because the region has seen the development and infilling of a series of earlier marine basins, by downwarping of the earth's crust or by tectonic activity involving the depression of a graben or valley between two more or less parallel fault lines. These basins and grabens were then infilled with sediments derived from upstanding land areas. Gradually heat and pressure has consolidated the loosely packed sediments, forming sandstones and shales, and in sampling these the drill returns evidence of fossil plant and animal remains which can be used to date the original sedimentary accumulations.

In and around the British Isles the development of Devonian sedimentary basins reflects the interference between faulting and folding in the central European system further to the south and east of the region and in the Arctic–North Atlantic system further to the west. The rapid subsidence of the largely fault-bounded Midland Valley Graben, the Northumberland–Solway Basin, the Dublin Trough, and the Orcadian Basin (Fig. 3.2) is related to this interference, and sediments were being deposited in each of these areas throughout Devonian times. The same massive stresses also led to the development of the Great Glen Fault, and Storedvedt (1974) estimates that the area to the north of the fault moved westwards by some 100–300km in late Devonian to early Carboniferous times.

An extensive Devonian basin also occupied much of the North Sea and apparently linked the Orcadian Basin, the Midland Valley Graben, and the Northumberland Basin as well as other sea areas on the west Norwegian coast with the Cornwall–Rhenish Basin further to the south. Little is known, however, about the structural configuration of this basin because it is deeply buried under younger sediments. However, it appears that, during mid to late Devonian times sandstones were being deposited throughout the Orcadian and North Sea Basins and that these shallow water deposits graded into deeper water shales and some carbonate sediments towards the south. Some material was also being eroded from the Welsh and London–Brabant Massifs and was being deposited in the central seas between these two small land masses, again grading westwards into deeper water shales and carbonates as shown in Fig. 3.2.

Carboniferous palaeogeography

The Eurasian and North American plates continued to converge throughout the transition from the Devonian to the Carboniferous periods. Additionally the European and African plate systems began to collide in the early Carboniferous period. These movements led to massive compression of the earth's crust and to the uplift and folding of the so-called Variscan deformation front in the south of the region. This fold complex (Fig. 3.3) finally closed the Cornwall–Rhenish basin. However, sedimentation rates exceeded subsidence in the new Variscan Basin so that its depositional regime changed from one of deep-water sedimentation into a shallower and eventually continental environment. This process ultimately gave rise to the accumulation of up to 3,500m of coal measures in lagoons and swamps across the region.

The Welsh Massif and London–Brabant Massif highs were considerably reduced in size, but continued to supply terrestrial sediments into the east–west basin which paralleled the Variscan fold belt. Previous highs, such as the fault-controlled northern Pennines and Irish blocks now formed relatively small islands within this elongate sea. The majority of the British Seas region was therefore exposed to erosion, but coal-bearing strata, interbedded with shales and occasionally coarser sediments were being deposited in the Southern North Sea, the Irish Sea, and to the west of Ireland. The numbers in Fig. 3.3 indicate the best estimates of the thicknesses of sediment deposited in these regions at this time.

The late Carboniferous–early Permian fault system of north-west Europe strongly fragmented the framework of the Variscan fold belt so that, shortly after its consolidation, it ceased to play a major role as a tectonic and morphological unit. The new fault system was reactivated time and again, and played a pre-eminent role in the development of the post-Variscan basins which are now preserved in the rocks beneath the British Seas.

The end of the Carboniferous and beginning of the Permian epochs finally completed the consolidation of the great Pangean super-continent, which welded together the Eurasian, North American and African plates. However, this massive continental system began almost immediately to show the signs of instability which were ultimately to lead to its break up and to the opening of the North and South Atlantic Oceans. Although crustal failure along the new rift system was not achieved during Permian times the way was being paved for the Jurassic, Cretaceous, and Early Cenozoic break-up of the Pangean super-continent.

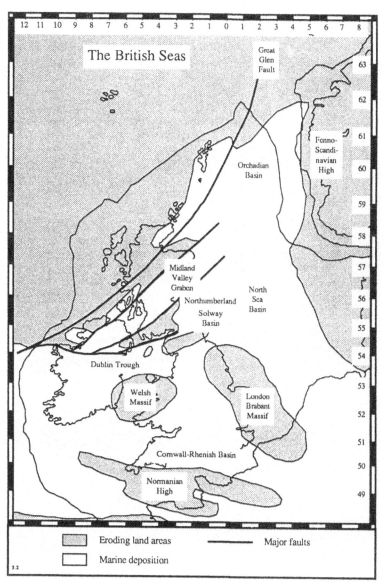

Figure 3.3 Late Carboniferous palaeogeography.

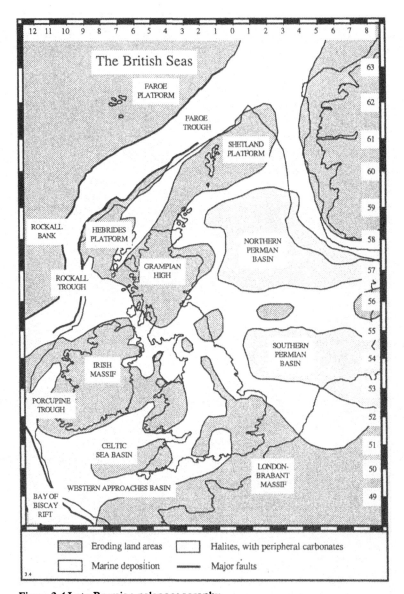

Figure 3.4 Late Permian palaeogeography.

Permian palaeogeography

During the early Permian, the continued uplift of the Variscan fold belt elevated much of the region well above sea-level. Within the fold belt a number of intra-montane (between highs) basins were trapped and continued to subside, and some were undoubtedly invaded by the Arctic Seas from the north and the proto-Atlantic Ocean from the west (Lorenz and Nicholls, 1976). Ziegler (1982) offers an early Permian palaeogeographical reconstruction showing a large Southern Permian Basin and a Northern Permian Basin to the east of the British Isles which are separated by a relatively narrow Mid-North Sea High. Smaller basins, perhaps in the Celtic Sea, the Western Approaches, and the Porcupine Trough are also indicated, but the Southern Permian Basin appears to have been a centre of deposition with some 15m of continental, as opposed to marine, sediments being deposited during these increasingly arid times. These beds were to form the Rotliegend Series red beds and halites which became of particular interest to the hydrocarbon industry in modern times (Chapter 9).

The opening of the North Atlantic now defines the present day western boundaries of the British Seas. It is not surprising, therefore, that much of the current discussion on the evolution of the region involves the dating of the earliest rifting in the Rockall–Faroe and Bay of Biscay areas, with one argument placing this in the early Permian, whilst another prefers a later date. There is little evidence either way, but the subsequent subsidence pattern is suggestive of a long-standing crustal extension which persisted throughout Permian times. Certainly rifting in late Permian times, which may have been accompanied by sea-level rise, led to the opening of the seaway between Norway and Greenland through which seas from the Arctic transgressed into the Northern and Southern Permian Basins of the North Sea (Pattison *et al.*, 1973).

In both the Southern and Northern Permian Basins, the advancing sea formed what is known as the Zechstein Transgression. The advance was seemingly very rapid, if not catastrophic, as is evidenced by the marine reworking of continental sediments in both basins. The shoreline moved landwards on all margins and some 1,000–2,000m of Zechstein sediments quickly accumulated in the North Sea (Fig. 3.4). Evidence suggests that the two basins also developed coastal reefs and carbonate deposits at times of high sea-level, and there was a corresponding sediment starvation in deeper water. By the end of the Permian, the Variscan fold belts had collapsed, and the environment is represented by rapid erosion of continental sediments into subsiding basins in the east of the region, with the development of rift valley

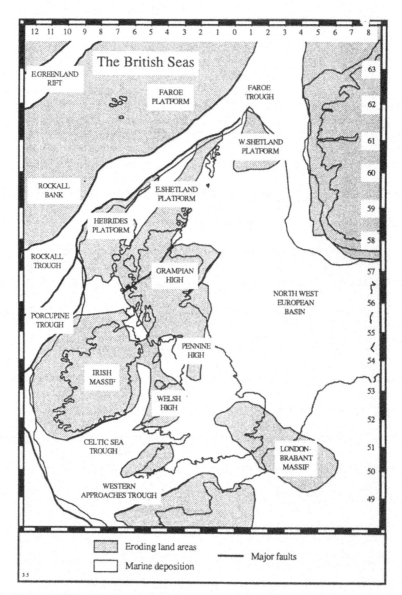

Figure 3.5 Triassic palaeogeography.

systems marking the proto-Atlantic Ocean to the west.

Triassic palaeogeography

The Triassic period in this region marks the onset of great rift system movements (Fig. 3.5) which are continuing to the present day, and began the opening of the Atlantic Oceans. The period commenced with a withdrawal or regression of the Zechstein Seas and a return to a continental depositional regime in both the North and South Permian Basins. The tensional elements were developing the Rockall–Faroe and the Bay of Biscay Rifts in the west and the Viking Central Graben in the east. The Triassic was then marked by a gradual global sea-level rise (Vail *et al.*, 1977), resulting in an extension of the Permian Basin margins. In the North Sea this resulted in the final submergence of the Mid-North Sea High (Fig. 3.4) and the formation of a large basin extending from the northern North Sea to southern Poland. The so called Muschelkalk Transgression extended from the south into this North-West European Basin and into the Bay of Biscay Rift. The fine grained Muschelkalk and succeeding Keuper marl sediments were thus deposited throughout large areas of the region, laying in places some 3000m of Triassic sediment.

Subsidence of the Rockall–Faroe Rift and the adjacent West Shetlands and Minches grabens continued apace, with some 4,000m of Permo-Triassic strata being deposited. Further south, opening of the Bay of Biscay Rift was accompanied by deposition in the subsiding Porcupine, Celtic Sea, Bristol Channel, and Western Approaches troughs of more than 2,000m of Triassic sediments.

Ziegler (1982) summarizes the Triassic by noting that the development of the region was entirely controlled by crustal extensions forming basins into which considerable thicknesses of sediments were deposited.

Jurassic palaeogeography

During the Jurassic period the proto-Atlantic Ocean came into evidence as one of the major driving elements governing the disintegration of the Pangean super-continent (Fig. 3.1).

The early Jurassic was probably associated with a temporary lowering of global sea-level and elevation of the existing land masses. The products of the erosion of these land areas are preserved as early Jurassic sedimentary deposits in the intervening lows. An irregular sea-level rise then produced a cyclical deposition of Jurassic sedi-

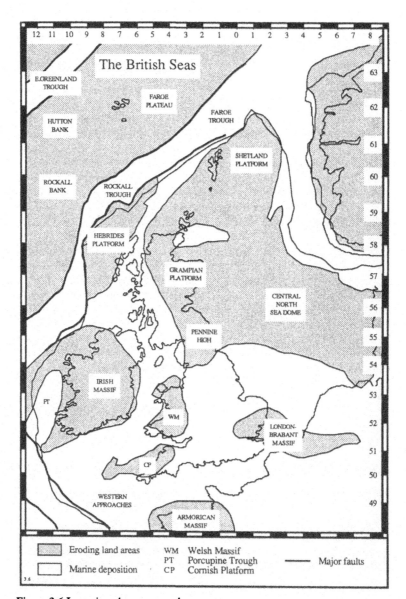

Figure 3.6 Jurassic palaeogeography.

ments which is known as the Lias. These are preserved throughout the region as shallow water shales and carbonates which generally contain only minor amounts of silt and sand. These were later to become the hydrocarbon source sediments in the Paris Basin, as well as in the North Sea and the Danish Trough. Liassic sedimentation rates generally kept pace more or less with subsidence, and the deposits were therefore fully preserved in the various basins, which still appeared to largely reflect their Triassic outlines.

At the transition from the early to the middle Jurassic, however, a major rifting phase affected the North-West European region and caused considerable palaeogeographic changes in the basins. In the North Sea, this activity led the central dome to be deeply truncated and shed sediments into the surrounding sea areas. Some of these deposits are represented by the regressive–transgressive sands of the Brent Group which were later to be of importance to the oil industry. In the Celtic Sea, Bristol Channel, and Western Approaches shallow marine shales, carbonates, and minor sands were deposited on top of the Lias, whilst, further north, continental clays and minor sands were being deposited in the Porcupine Trough.

The North Sea dome began to subside in the late Jurassic which, combined with generally rising sea-levels, led to a further transgression and the seaway was again opened throughout the basin. The Viking and the central grabens then developed into the major depositional feature of the region. To the west the Porcupine Trough, the Celtic Sea, Bristol Channel, and Western Approaches grabens also continued to subside and crustal extensions developed which were to lead to the opening of the North Atlantic.

Thus, the palaeogeography in late Jurassic times (Fig. 3.6) shows a whole series of largely rift and graben controlled highs separated by relatively shallow and infilling sea areas. The marine sedimentation to the south is largely of shales and carbonates, whilst the calcium-rich sediments are replaced further north by shales and some local, coastal sands.

Cretaceous palaeogeography

Further phases of rifting in early Cretaceous times were accompanied by significant eustatic sea-level drops that are expressed by a regional regression and relative elevation of the pre-existing high areas. The results were an emergence of considerable land areas and temporary restriction of sedimentation to the deeper parts of the North Sea Rift and the western basins. These movements signalled the onset of the seafloor spreading along the Bay of Biscay and Rock-

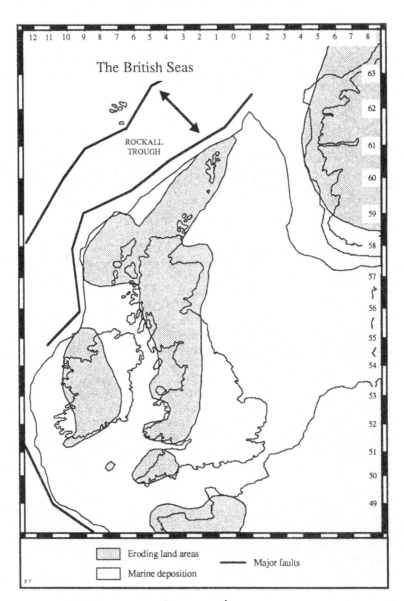

Figure 3.7 Late Cretaceous palaeogeography.

all Trough. Although considerable thicknesses of sediment were deposited in the North Sea and the Porcupine Trough in the early Cretaceous, it is probable that the palaeogeography had altered little from Jurassic times.

In contrast to this gentle introduction to the period, the late Cretaceous is marked by, on the one hand, the progressive opening of the North Atlantic and the seafloor spreading in the Norwegian–Greenland Sea, and on the other the onset of the Alpine folding further south as a continuing consequence of the collision of the African and Eurasian plates.

The first result of these events in the region was a significant rise in global sea-levels (Pitman, 1978), reaching a maximum high stand of some 110–300m or more above the present level. In the British Seas this gave rise to the late Cretaceous transgression, which in turn drastically reduced the influx of sands and silts from the land masses and resulted in the prevalence of clear-water conditions and the deposition of the chalk (Fig. 3.7). The bulk of these chalk deposits are composed of small animals called coccoliths, although in the northeast of the region a deep water, shaley, and marly facies is developed with a total thickness of more than 2000m. Also, the thickness of the chalk to the south-west indicates a gentle oceanward tilt of the whole of the Celtic Sea–Western Approaches shelf as a consequence of increased seafloor spreading rates in both the Rockall Trough and the Bay of Biscay.

Thus events at the end of the Cretaceous period were severely influenced by the two different processes, the opening of the North Atlantic and the effects of the Alpine fold system.

Cenozoic palaeogeography

The most recent era of geological time is the Cenozoic, which began about 63 million years ago and has continued to the present day. Not surprisingly, the relative abundance of deposits, the profusion of remains from animals and plants, and the tendency for Cenozoic sediments to be found closer to the present day land surface have all contributed to a more detailed and a more sophisticated reconstruction of this recent earth history. The era is subdivided into three periods: the Palaeogene, the Neogene, and the Quaternary as shown in Table 3.2, although the Palaeogene and Neogene are often referred to as simply the Tertiary.

The Cenozoic basins of north-west Europe developed in response to both the widening Atlantic to the west, and to the effects of the Alpine movements further south. The subsidence and infilling of the

41

Figure 3.8 Cenozoic palaeogeography.

Table 3.2 Stratigraphic divisions of the Cenozoic

Period	Epoch	Duration
Quaternary	Holocene	about 20,000 years
	Pleistocene	about 2 million years
Neogene	Pliocene	about 5 million years
	Miocene	19 million years
Palaeogene	Oligocene	12 million years
	Eocene	16 million years
	Palaeocene	10 million years

North Sea Basin, the Atlantic shelf areas and the Porcupine Trough (Fig. 3.8) were a direct response to the seafloor spreading of the, by now, well-defined western margins of the shelf. At the same time, the foredeep basins associated with the Alpine movements showed initial subsidence which was followed by intense folding and thrust developments. Throughout the Cenozoic then, the present-day coastlines were developing and the present-day regions of deposition within the British Seas were becoming well-defined. However, it is the marine processes during the latest 2 million years of geological history which has finally sculptured the main bathymetric features described in the preceding chapter. These processes are closely related to relative and absolute rises and falls in sea-level due to a series of global ice-ages. These changes are described in the following section.

Quaternary palaeogeography

Although the effects of changing sea-level patterns on sedimentation in the British Seas have been mentioned in the preceding sections, the record is incomplete because the rocks are often disturbed through tectonic activity, sediment erosion and transportation, changes in the overall shape of the earth, and because of other, unrecorded climatic events. Vail *et al.* (1977) have reconstructed a 600 million year sea-level curve (Fig. 3.9) which shows a number of rises and falls of sea-level over a 600m vertical range. The picture becomes much clearer, however, when the most recent events of the geological past are examined. The expansion and contraction of the Quaternary ice-sheets have clearly led to rises and falls in sea-level which had an overriding influence on the shape of the present day north-west European shelf. Figure 3.9(b) and (c) show estimates of sea-level during the past 150,000 years. The rise in sea-level due to ice melting is called eustatic, as opposed to the isostatic changes which are due to

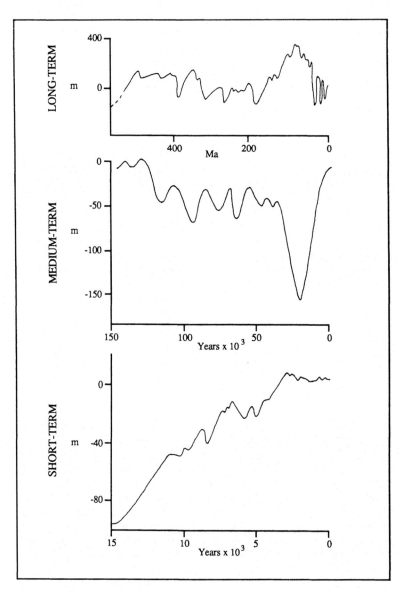

Figure 3.9 Sea-level rise.
Source: Carter (1988).

uplift or subsidence of the land masses. The most recent, the Ho-
locene, rise has been extensively studied (Bloom, 1977) and it is now
clear that the search for a 'global' sea-level curve is probably fruitless
and attention has now focused on determining local peculiarities of
relative uplift and eustatic rises and falls.

However, virtually all of the curves show an extreme sea-level low
of between 100 and 150m around 20,000 years before present (BP). In
the British Seas this meant that the whole of the North Sea and English Channel were exposed and had great fluvial drainage systems
extending northwards from the Rhine, Thames, and Humber and
westwards from the Seine (Hamilton and Smith, 1972, and Chapter
2). All curves also show a strong rise between 15,000 and 8,000–
6,000 BP. In the British Seas the rate of sea-level rise outstripped
glacio- and hydrostatic adjustments, producing a major marine transgression and flooding in all but the most elevated regions of the
southern North Sea. Sea-level has remained relatively constant at or
about its present day height for the last 5,000 years on a worldwide
basis, but there is some evidence that this level was not achieved until
3,500–4,000 years BP in Europe (Kidson, 1982). Recent debate over
levels in, for example, the Irish Sea is still continuing (Kidson, 1982;
Carter, 1988).

The results of these transgressions are considerable thicknesses of
Cenozoic sediments in the British Seas. In the North Sea, for
example, Caston (1979) suggests that up to 1,000m of Quaternary
sediments infill a broad basin which is centred around 57°N and runs
the whole length of the North Sea from Germany to beyond 62°N.
Pantin and Evans (1984) provide an excellent review of Quaternary
sedimentation patterns in the Celtic Sea and utilize a numerical
model to show that the Melville Bank (Chapter 2) was probably
formed by tidal currents at a time of lowered sea-level. Again Bowen
(1978) and Pantin (1989) consider the likely extent of the Quaternary
ice-sheets in the British Seas, and their results are shown in Figure
3.8. The surface sediments of the British Seas, therefore, represent a
complex amalgam of locally eroded and river derived material which
has been reworked by repeated marine inundations, by glacial advances, and by meltwater channels associated with the ice-sheets. These
sediments are now being eroded, transported, and re-deposited by
the present-day action of waves and tides, and we must consider these
oceanographic processes in the following chapters before returning
to the modern-day sedimentological regime in Chapter 7.

The oceanography

References

Bloom, A.L., 1977. *Atlas of Sea-Level Curves*. IGCP-200, Cornell University Press, New York.

Bowen, D.Q., 1978. *Quaternary Geology*. Pergamon Press, Oxford.

Carter, R.W.G., 1988. *Coastal Environments*. Academic Press, London, 617 pp.

Caston, V.N.D., 1979. The Quaternary sediments of the North Sea. In: Banner, F.T., Collins, M.B. and Massie K.S. (Eds) *The North-West European Shelf Seas: The Sea Bed and Sea in Motion*, Vol. I, *Geology and Sedimentology*. Elsevier, Amsterdam, pp. 195–270.

Eagar, R.M.C., 1976. *The Geological Column*, 5th Edn. Manchester Museum Publication.

Hamilton, D. and A.J. Smith, 1972. The origin and sedimentary history of the Hurd Deep, English Channel, with additional notes on other deeps in the western English Channel. *Mem. Bur. Rech. Geol. Min.*, 79, 59–78.

Kidson, C., 1982. *Quaternary Sci. Rev.*, 1, 121–151.

Lorenz, V. and I.A. Nicholls, 1976. The Permocarboniferous Basin and Range Province of Europe, an Application of Plate Tectonics. In: Falke, H. (Ed.) *The Continental Permian in Central, West and Southern Europe*. NATO Advanced Study Institutes Series C, Math. Phys. Sci., 22, 313–342.

Pantin, H.M., 1989. *Report on the 1:1,000,000 Seabed Sediments Around the United Kingdom Sheets (North and South)*. British Geological Survey.

Pantin, H.M. and C.D.R. Evans, 1984. The Quaternary history of the central and southwestern Celtic Sea. *Mar. Geol.*, 57, 259–293.

Pattison, J., D.B. Smith and I. Warrington, 1973. A review of late Permian and early Triassic biostratigraphy in the British Isles. In: Logan, A. and Hills, L.V. (Eds) *The Permian and Their Mutual Boundary*. *Can. Soc. Petrol. Geol. Mem.*, 2, 220–260.

Pitman, W.C., 1978. Relationship between eustacy and stratigraphic sequences on passive margins. *Geol. Soc. Am. Bull.*, 83, 619–646.

Storedvedt, K.M., 1974. A possible large scale sinsitral displacement of the Great Glen Fault in Scotland. *Geol. Mag.*, 111, 23–30

Vail, P.R., R.M. Mitchum, R.G. Todd, J.M. Widmeir, S. Thompson, J.B. Sangree, J.N. Bubb, and W.G. Hatfield, 1977. Seismic stratigraphy and global changes of sealevel. In: Payton, C.E. (Ed.) *Seismic Stratigraphy, Application to Hydrocarbon Exploration. Am. Ass. Petrol. Geol. Mem.*, 26, 42–212.

Ziegler, P.A., 1982. Evolution of sedimentary basins in North-West Europe. In: Illing, L.V. and Hobson, G.D. (Eds) *The Petroleum Geology of the Continental Shelf of North-West Europe*. Academic Press, London, pp. 3–39.

Chapter four

The wave regime

Waves, whether generated by trade winds far out in the deep oceans, or by local storm events in coastal waters, have a number of important influences on the oceanography and resources of the British Seas. In the first case long waves with periods of 15–20s are capable of inducing currents down to the seabed and are thus able to mix otherwise stratified water structures (Chapter 6) and to transport the seabed sediments (Chapter 7). Additionally, waves and particularly severe storm conditions can, on the one hand, affect or even suspend maritime industrial work, or on the other hand supply enormous quantities of energy to suitable wave-power stations (Chapter 12). The sedimentologist must, therefore, allow for wave action in his interpretation of the seabed regime, whilst the businessman planning the introduction of a new ferry link may well find his optimistic profit forecasts are reduced to the breakeven point or worse, when adverse conditions halt both the service and the ticket revenue. In this chapter we examine the techniques of predicting long-term wave conditions in shelf waters in order to overcome some of the problems mentioned above. We exemplify these techniques with the seas around the British Isles. The presentation is based largely on Draper (1979), with some more specific references included in the text. The additional literature on ocean waves is very extensive and the interested reader should refer to Hardisty (1990) for wave recording, Wiegel (1964) for the mathematics of the wave equations, and to Hardisty (forthcoming) for a fuller treatment of wave-dominated geomorphological and sedimentological processes.

Draper (1980) reviews the development of wave recording and suggests that some of the earliest data were collected by Thomas Stevenson, the engineer and builder of lighthouses and harbours in Scotland who was 'interested in sea waves because they tried to demolish his carefully constructed structures'. Stevenson instructed Mr William Middlemiss, his resident engineer at Lybstor in Caithness, to

Plate 4 Waves transfer energy from the deep ocean to the shelf

note the height of the waves every day in 1852 and he published the results in Stevenson (1857). Stevenson also produced a wave prediction formula, $H = 1.5\sqrt{F}$, where H is the design wave height in feet and F is the fetch in miles. He must have been an astute observer, for the most modern observations are in general accord with his formula. For example, the modern estimate of the 50-year wave (see below) in the eastern North Atlantic is about 30m whereas Stevenson's formula with a fetch of 4,000 miles predicts a value of 95 feet! The earliest electronic records of coastal waves were obtained in 1947 off Perranporth in Cornwall (Darbyshire, 1962) and the first offshore measurements were obtained at the Morecambe Bay Light Vessel in 1956–7 (Draper, 1968).

Wave prediction

Waves are generated when wind, blowing over the surface of the open ocean, transfers energy from the atmosphere to the water. The result is a series of more or less parallel crests and troughs which progress in a generally downwind direction across the sea surface. The profile of a deep-water wave is shown in Figure 4.1(a) and from this we define six wave parameters.

1. The significant wave height, H_s, is the average value of the vertical distance between the crest and the trough of the highest one-third of all waves present.
2. The maximum wave height in a given time interval $H_{max(time)}$ is the maximum wave height which is likely to occur in, say, 10 minutes, $H_{max(10min)}$ or, as is used for engineering purposes, in 50 years, $H_{max(50years)}$
3. Wave period, T, is the time which elapses between the passage of one crest (or trough) and the next crest (or trough), or one particular point on the wave profile and the same point on the next wave.
4. Zero crossing period, T_z, is the average time interval over which the water surface successively passes upwards through the mean water level. For typical wave conditions it can range between 2 and 20s.
5. Significant wave period, T_s, is the mean period of the waves involved in the significant wave height definition.
6. Wave length, L, is the horizontal distance from a particular point (for example, a crest) on one wave to the same point on the next wave. It is difficult to measure the wave length directly, but in deep water it can be calculated from the wave period

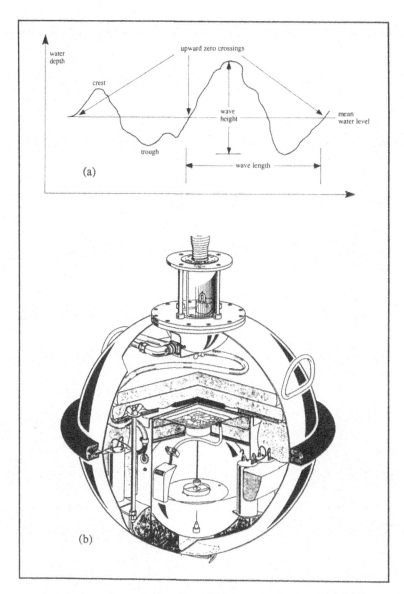

Figure 4.1 Diagrammatic explanation of (a) wave parameters, and (b) the Wave-rider buoy.

by the relationship $L = 1.56T^2$ where T is in seconds and L is in metres.

These wave parameters are determined for the UK continental shelf waters either by direct measurements or by prediction from known wind conditions. Each procedure will be dealt with separately before proceeding to present a synopsis of the results for the area.

Deep water wave measurements

In shallow, coastal waters sensors can be mounted on the seabed or on fixed structures at the shoreline to obtain long-term measurements of wave height and period. This is not possible further offshore, because structures are not fixed and the effects of waves are generally attenuated far above the seabed. Instead, recourse must be made to sensors mounted on ships or moored buoys and compensation must then be made for the movements of the platform before the wave heights can be determined.

The 'Shipborne' wave recorder was the first instrument to be routinely deployed in the British Seas. It was developed at the British Institute of Oceanographic Sciences and has been extensively deployed in UK waters over the last 20 years. The system consists of a hull-mounted pressure transducer which measures the depth of water above a pressure port, but the pitching and rolling means that this signal is not simply the wave record. To compensate for ship movement a vertical accelerometer is mounted close to the pressure port which computes the position of the port with respect to the still water level, so that the difference between the two signals gives the true wave height. More recently 'Waverider' buoys have replaced 'Shipborne' systems for routine offshore work (Fig. 4.1(b)). The Waverider has a central gimbal which measures the acceleration of the buoy as it floats on the water surface. Since the first integral of acceleration is velocity and the first integral of velocity is in turn displacement, then the device measures the displacement of the water surface by calculating the double integral of the acceleration measurements to give the surface profile of the waves. The data is usually transmitted to a shore-based recorder through the aerial on the top of the buoy. Most wave data in the British Seas are collected by the Marine Information and Advisory Service (recently retitled the British Oceanographic Data Centre) and the present extent of reliable offshore measurements is shown in Fig. 4.2.

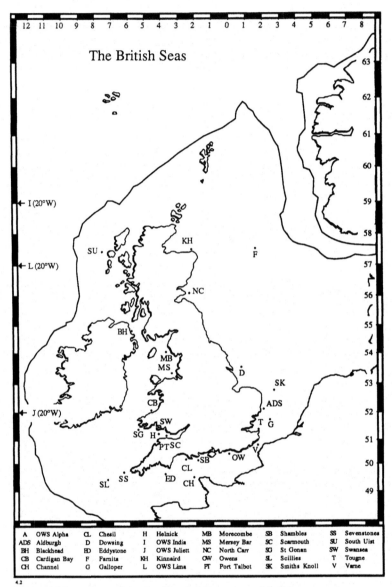

Figure 4.2 Locations of the Institute of Oceanographic Sciences wave measurement sites at shore stations, light ships, and ocean weather ships.

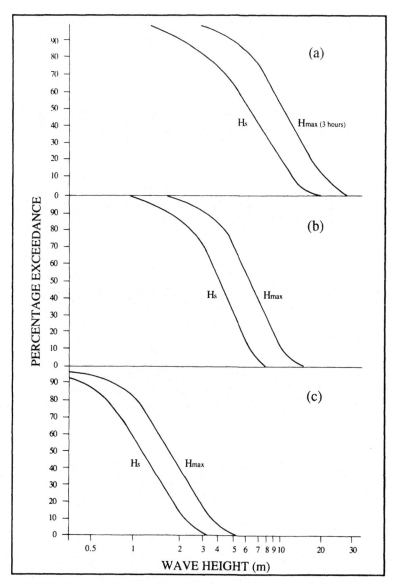

Figure 4.3 Comparison of the maximum and significant wave height at (a)
Ocean Weather Ship *India* (59N, 19W), (b) Sevenstones off Land's End and
(c) the Varne Light Ship in the Dover Straits.

Source: Draper (1979).

Wave data presentation

One of the many problems of wave recording involves the presentation of the results in a meaningful and practical form. Most usefully, annual results are now reported in a series of diagrams using the Tucker Draper Method (TDM). The results are summarized by five different plots in the TDM:

1. Percentage exceedance diagrams for the significant and maximum wave heights as shown in Fig. 4.3. These represent the cumulative frequency of each wave height, for both H_{max} and H_s.
2. Zero-crossing diagrams which represent a histogram of the distribution of wave periods throughout the year.
3. Scatter plots of T_z versus H_s as shown in Fig. 4.4(a). Since the wave length is related to the wave period, as we have seen, then these diagrams give an indication of the steepness of the waves.
4. Wave persistence diagrams as shown in Fig. 4.4(b). These represent the likely time interval for which particular wave heights are to be exceeded, and are thus useful for calculating the downtime of particular marine operations. This aspect of wave data analysis is discussed further at the end of the present chapter.
5. Extreme value predictions using the Weibull probability scale as shown in Fig. 4.4(c). It is useful, for engineering purposes, to estimate the highest wave which is likely to be experienced in a given, say 50-year, period. This estimate is achieved by plotting the percentage exceedance data on a Weibull probability scale. Waves, like other natural phenomena such as wind speeds, tend to approximate to a Weibull distribution, and therefore to plot as a straight line on such a scale. The results from 1 or 2 years of recording can therefore be extrapolated to much longer periods, in order to determine extreme value wave heights. In Fig. 4.4(c), it can be seen that the 50-year wave for the Varne site will be about 25 feet (8m) high.

Deep water wave prediction

When wave measurements are not available it is sometimes necessary to predict the wave parameters from given wind conditions. In UK waters a graphical method which is based upon measurements of waves on the shelf (Darbyshire, 1959a, b) has been developed to facilitate these predictions. The graph presented by Driver (1980) has

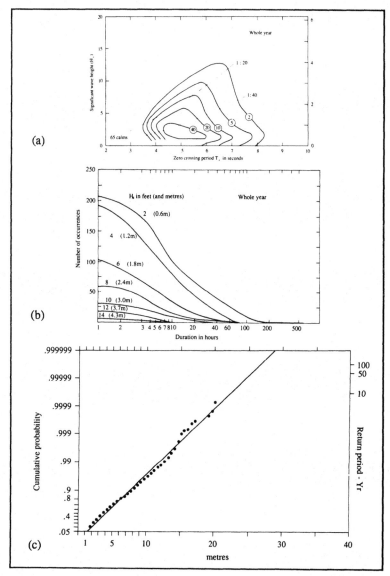

Figure 4.4 Presentation of wave data in (a) period and height scattergram, (b) exceedance diagram, and (c) recurrence probability diagram. All data are for the Varne Light Vessel.

Source: Draper (1979).

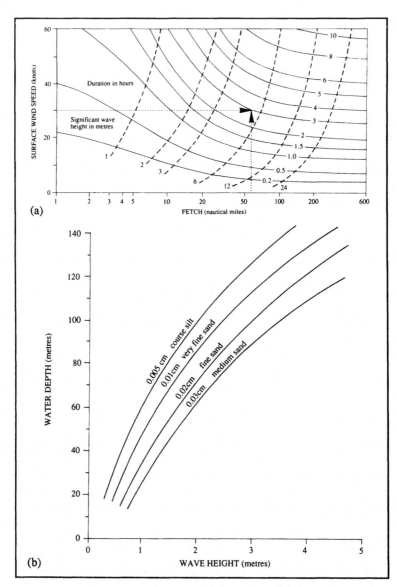

Figure 4.5 (a) Wave height forecasting graphs and (b) threshold of seabed sediment transport for a wave period of 15s.

been converted into metric units and is shown in Fig. 4.5(a), although the fetch length is still given in the more usual nautical miles. Although the mechanisms of wave generation are not fully understood, the relationship which is implied by the diagram, that wave height increases with both wind strength and fetch length, does achieve an acceptable degree of accuracy for most purposes. For example, using Fig. 4.5(a), with a wind speed of 30 knots blowing over a 60 nautical mile fetch for 12 hours, we enter the graph from the left at 30 knots and proceed horizontally until either the fetch limit line (vertical from 60 nautical miles) or the duration limit line (broken line labelled 12h) is reached and the wave height, 3m, is read off.

Wave climate

There is little difference between the waves which occur in the deep Atlantic Ocean to the north-west of the UK and those which occur to the south-west. The wave climates from Ocean Weather Ships *India* and *Juliet* (Fig. 4.2) are virtually indistinguishable from one another (Draper and Squire, 1967; Draper and Whitaker, 1965; Draper, 1979). Deep water in this context implies a depth of 200m or more, beyond the edge of the continental shelf where the waves are, for all practical purposes, unaffected by the seabed. The maximum energy of severe ocean storm waves in these regions occurs at a period of about 16s, and has been measured with Shipborne wave recorders on Ocean Weather Ships *India*, *Juliet* and *Lima*. The quietest open sea areas are those to the south-east of England, and there are only small differences in the wave climate from east of the Isle of Wight to the southern North Sea east of Lowestoft (Draper, 1979). The change in wave severity can be demonstrated by comparing the significant winter wave height along a line from the west to the east of the region as shown in Fig. 4.3. The significant wave height at the ocean weather stations exceeds 4.0m for 50 per cent of the time. At the most exposed light vessel station, Sevenstones, the equivalent height is 2.8m, and at the Varne Light Vessel in the eastern English Channel it is under 0.9m. In the summer, data from the three stations are still in the same relative proportions although the wave heights are about half of the winter value. The most commonly occurring wave periods in the Atlantic are around 10s in the winter, falling to about 8s in the summer; whilst at the coastal stations the values vary only minimally throughout the year. This suggests that the longer-period wave energy, resulting from the stronger winter winds, is lost through the effect of seabed friction with the result that the wave steepness in

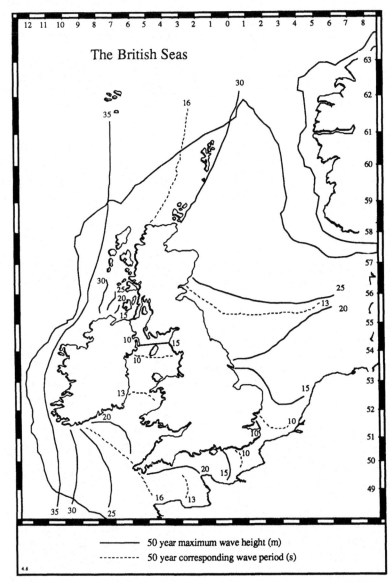

Figure 4.6 The 50-year wave height and period distributions for the British Seas.

Source: Lee and Ramster (1981).

these coastal areas is higher, especially in the winter.

Figure 4.6 provides a convenient method of comparing the severity of wave conditions in different areas across the shelf. The chart is based on Lee and Ramster (1981) and has been compiled from two main sources of wave data. The one is the determination of extreme waves by the wave forecasting method using estimates of the 50-year extreme winds prepared by the British Meteorological Office. It assumes that the storm responsible for the waves will last in its fully developed state for 12 hours. The other source relies on the extrapolation of long series of instrumentally measured wave data, mainly from the light vessels and ocean weather ships shown in Fig. 4.2.

The figure shows the best available estimates from region to region of the heights of the highest individual waves which are likely to occur in any 50-year interval, and also the mean zero-crossing period of the waves at that time. Since the wave length in deep water is related to the wave period by $L = 1.56T^2$ then the figure provides an idea not only of the height of the worst wave which might be encountered by a ship or an offshore structure, but also of the distance between wave crests. It should be noted that because of the statistical nature of sea waves the predicted wave height may not be achieved, or it may be exceeded on more than one occasion in any 50-year interval. In general the wave height decreases from the open ocean values due to seabed friction as the waves propagate eastwards along the English Channel into the Irish Sea or southwards, down the North Sea. The more exposed Celtic Sea and North-western Approaches suffer the full severity of the deep ocean waves.

Wave effects on seabed sediments

From information on the wave climate at the sea surface it is possible to calculate the magnitude of wave disturbance at the seabed. The water at the surface moves in a circular orbit with a diameter equal to the wave height; at the seabed the vertical component is suppressed and the water particles move only horizontally, with zero speeds as the particles reverse direction at the end of each orbit. Komar and Miller (1975) give equations for the threshold of sediment transport as a function of the wave parameters and the water depth as shown in Fig. 4.5(b). The results are for waves with a period of about 15s, which is similar to those described above for storm conditions in the British Seas. It is apparent that such waves with, for example, heights of more than 4m will be competent to transport sand-sized sediments in water depths of up to 100m, and silts in depths of 150m or more. Since these heights are frequently exceeded (Fig. 4.3), it is likely that

The oceanography

surface waves are responsible for disturbing the seabed sediments across all of the shelf during severe storm conditions. This result will be carried forward to the discussion of seabed sediments which is presented in Chapter 7.

Wave effects on shipping

The safe and successful operation of vessels, buoys, nets, pipeline barges, platforms, dredgers and so forth depends upon the local wave conditions. Estimates of likely downtime due to rough weather must therefore be incorporated in any marine contract and can be predicted from a knowledge of the wave regime. Wiegel (1964) summarizes some results and his listings are shown in Table 4.1. The likely downtime can then be calculated from the type of persistence diagram shown in Fig. 4.4.

Table 4.1 Generalized performance data for marine operations (after Wiegel, 1964)

Type of Operation	Maximum Wave Heights (m) for:		
	Safe Operation	Marginal Operation	Dangerous Operation
Deep-sea tug	1	2	>2
Crew boat (30m LWL)			
Underway	3	5	>5
Crew transfer	1	2	>2
Buoy laying	0.7	1	>1
Craning	1	2	>2
Gravity meter exploration	1.5	2	>2
Seismic surveys	2	3	>3
Amphibious aircraft	0.3	0.6	>0.6

Source: Wiegel (1964)

References

Darbyshire, J., 1959a. A further investigation of wind generated waves. *Dtsch. Hydrogr. Z.*, 12(1).

Darbyshire, J., 1959b. The spectra of coastal waves. *Dtsch. Hydrogr. Z.*, 12(4).

Darbyshire, M., 1962. Wave measurements made by the National Institute of Oceanography. *Marine Observer*, 32, 32–40.

Draper, L., 1967. Wave activity at the seabed around northwestern Europe, *Mar.Geol.*, 5, 133–140.

Draper, L., 1968. *Waves at the Morecombe Bay Light Vessel*. National Institute of Oceanography, Report A32.

Draper, L., 1979, Wave climatology of the U.K. Continental Shelf. In Banner, F.T., Collins M.B. and Massie, K.S. (Eds) *Northwest European Shelf Seas: The Sea Bed and Sea in Motion*, Vol. I, *Geology and Sedimentology*. Elsevier, Amsterdam, pp. 353–368

Draper, L., 1980. The reliability of sea wave data. In: Count, B. (Ed.) *Power from the Seas*. Academic Press, London.

Draper, L., and Squire, E.M., 1967. Waves at ocean weather ship station 'India', *R. Inst. Nav. Archit (Lon) Trans.*, 109, 85–93

Draper, L. and M.A.B. Whitaker, 1965. Waves at ocean weather ship station 'Juliet', *Dtsch. Hydrogr. Z.*, 18(1), 25–30

Driver, J.S., 1980. A guide to sea wave recording. Institute of Oceanographic Sciences, unpublished report 103, 51 pp.

Hardisty, J., 1990. An introduction to wave recording for coastal geomorphologists. *Tech. Bull. Brit. Geomorph. Res. Group*, 39, 76 pp.

Hardisty, J., forthcoming. *Beaches: Form and Process*. Unwin-Hyman, London.

Komar, P.D. and M.C. Miller, 1975. On the comparison between the threshold of sediment motion under waves and unidirectional currents with a discussion of the practical evaluation of the threshold. *J. Sediment. Petrol.*, 45, 362–7

Lee, A.J. and J.W. Ramster, 1981. *Atlas of the Seas Around the British Isles*. MAFF, Lowestoft.

Stevenson,T., 1857. *The Design and Construction of Harbours, a Treatise on Maritime Engineering*, 2nd Edn. Edinburgh.

Wiegel, R.L., 1964. *Oceanographical Engineering*. Prentice-Hall, New Jersey, 532 pp.

Chapter five

The tidal regime

The aim of this chapter is to describe the movements of tidal waters in the British Seas. These water movements are both vertical and horizontal: vertical displacements of the water surface with approximately twice-daily period are familiar to all shore-based observers, but to the mariner the twice-daily variations in the horizontal movements, called tidal streams, are equally important. Although there is a long history of observations of tides, both for practical and for scientific purposes, the relatively recent advent of electronic technology has enabled a systematic study with a detail which would previously have been impossible. Variations in sea-level and currents may now be measured and automatically recorded over periods of months at any position on the continental shelf; furthermore, these observations can be processed and analysed by computer both quickly and accurately. Not only are these detailed descriptions of tides now possible, but they are also necessary, in order to meet the greater accuracy required by mariners, coastal engineers, and others who plan how the characteristics and resources of the sea may best be protected and utilized.

This chapter also provides the background information for two other chapters in the book. From the oceanographic point of view, an understanding of the tidal currents in the British Seas will be required to appreciate the great seabed sand transport paths which are described in Chapter 7. From the point of view of resource utilization, the tides provide one of the two marine energy sources which are discussed in Chapter 12.

The chapter is based on Huntley (1979), Howarth & Pugh (1983) and Pugh (1987) together with the more specific references which are included in the text. We define a tide as a *periodic vertical or horizontal movement which has a coherent amplitude and phase relationship to some periodic geophysical force*. In our case we consider the movement of water, but movements of the atmosphere and of solid earth

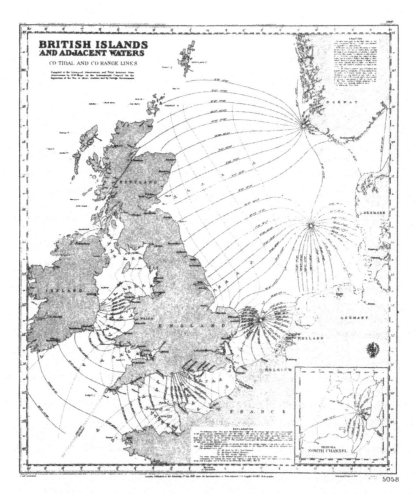

Plate 5 This 1937 M tidal chart showing cotidal and corange lines was one of the first attempts to depict the details of tidal forcing across the offshore regions (Crown copyright – reproduced from Admiralty Charts 1937 with the permission of the controller of Her Majesty's Stationery Office)

may also be tidal. The dominant geophysical force is due to the variation in the gravitational field at the surface of the earth, caused by the regular movements of the earth–moon and earth–sun systems. Movements due to the sun and the moon are termed 'gravitational tides' to distinguish them from movements induced by regular meteorological forcing. The latter are termed 'meteorological tides' because they occur at periods directly linked to the solar day.

Measurements of tidal elevations

A brief history

The earliest recorded observations of tides on the north-west European continental shelf seem to have been made by Pytheas of Marseilles in about 320 BC. Pytheas made a voyage of discovery in which he circumnavigated the British Isles, and he reported seeing tides which rose to a height of 36m on the coast of Britain, a gross overestimate typical of tidal observations even as late as the seventeenth century. Pytheas also knew of the connection between the tides and the motion of the moon, and the details of this connection were slowly unfolded during the following centuries.

It was not until the thirteenth century that science began to move forward again and concentrated attempts were made to record and to explain the observed tides around British shores. Gerald of Wales (c.1146–c.1220) recorded observations of the tides in the Irish Sea, and the earliest British tide tables also date from this time. They forecast, in a very crude way, the times of high water at London Bridge and are credited to John Wallingford, Abbot of St Albans, who died in 1213.

Progress continued to be intermittent until, in the seventeenth century, Sir Robert Moray made tidal observations in the Western Isles of Scotland, from which he was able to deduce that the ebb and flood of the tide, and the monthly progression of high tides were both nearly sinusoidal.

In 1666, the Royal Society asked Moray to draw up instructions for making observations both of the rate of ebb and flood and the variation of tide height. Moray designed a completely new type of tide measuring instrument as part of a tidal observatory, which, he suggested, should be built at places with a large tidal range. Unfortunately, Moray's plans were too ambitious for their time, and for the next 160 years, although great progress was made in the theoretical understanding of tidal motion (in particular by the publication of Newton's *Principia* in 1687), tidal observations were still made by the

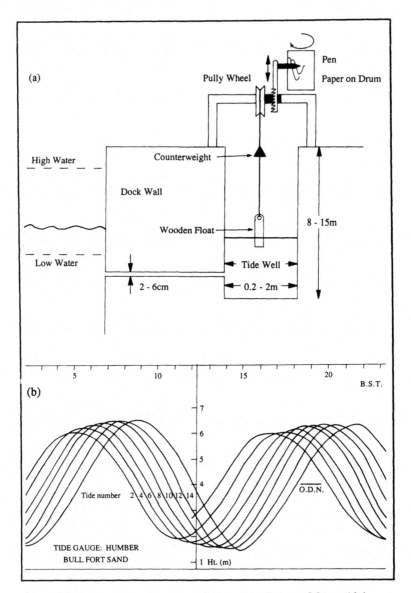

Figure 5.1 (a) Diagrammatic representation of the Palmer–Moray tidal stilling well and (b) typical tide curves from Bull Fort at the mouth of the Humber Estuary.

66

ancient method of estimating mean water level against a vertical tide staff.

Nevertheless, with increasing overseas trading in the early part of the nineteenth century, the need for accurate tide measurements became more pressing. Henry Palmer designed a self-recording tide gauge, based on the plans of Moray, and published his design in 1831. Palmer's tide gauge enabled the complete shape of the tide elevation curve to be recorded, and would operate unattended for long periods. The first gauge was set up in London in 1828 and was quickly followed by others at Sheerness, Portsmouth, Plymouth and Bristol. The first volume of the Admiralty Tide Tables was published in 1833 using results from the London, Sheerness, Plymouth and Portsmouth gauges. With these developments, the scientific analysis of tidal phenomena had at last begun.

Modern gauging stations

The Palmer–Moray tide gauge is still the most important instrument for measuring tidal elevation on the shelf seas. Its essential feature is a 'stilling well', a well reaching to below the level of low water. This well is connected to the sea by a narrow orifice which serves to dampen the influence of short period surface waves while allowing the well level to follow accurately the long period tidal movement. The level of water in the well is measured by a float, usually of wood, which is connected to a counterweight by a string running over a series of pulleys at the top of the well. The pulleys act through gearing to move a pen which records the tidal elevation on a cylindrical drum of paper rotated at a known speed by a clockwork motor. The basic components are shown diagrammatically in Fig. 5.1(a). Fig. 5.1(b) is a typical record, from the gauge at Bull Fort in the North Sea entrance to the Humber Estuary, and shows the variation in water level over a two-week period. The paper has been removed from a drum which rotated once every 24h, so that the predominantly semi-diurnal (twice-daily) tide, lagging by about 50min in every 24h, appears progressively displaced from day to day. The succession of tidal range from a large spring range to a much smaller neap range is also clearly represented on this record.

This gauge forms the pattern for most of the tide gauges now operating around the British Seas. Figure 5.2 shows the distribution of those tide gauges around the north-west European shelf sea coasts considered sufficiently reliable to be used by the UNESCO Permanent Services for Mean Sea Level, but there are many more tide wells in use by local and port authorities.

In deeper water, seabed-mounted pressure transducers have been

67

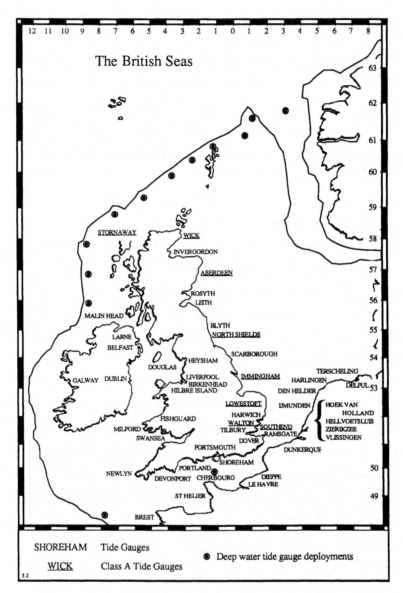

Figure 5.2 Distribution of permanent coastal tide gauges and short-term deep-water deployments around the UK.

deployed which have on-board recording devices and the progress of the tidal wave across the outer shelf has now been determined from a string of such measurements at the sites shown in Fig. 5.2.

Measurements of tidal currents

Recorded observations of tidal streams as opposed to tidal elevations on the shelf seas are numerous, but most are short-period (25 h) observations of surface or near surface currents, and are of very poor quality by comparison with the tidal elevation data. Nevertheless the accuracy is sufficient for most navigational purposes, and the important data are collected in atlases and tables of tidal streams. The data presented in these charts have been accumulated over the past century, and come from a variety of sources. Most commonly they are observations, made from survey ships, of the drift velocity of surface floats or of a pole, floating vertically with its top just at the water surface. Less commonly, observations from fixed lightships and even some from self-recording current meters, generally moored 10m below the surface, are used.

It is usual to plot 25h (i.e. two semi-diurnal periods) of current magnitudes and directions on a vector diagram, in order to eliminate spurious effects from the observations. The resulting plot will generally be in the form of an ellipse, with the direction of the tidal stream passing through every point of the compass during the course of a single tidal cycle; the best estimate of the tidal stream ellipse is then found by drawing a smooth curve through the observed points. Long-period measurements of tidal streams at different depths are now being made with vertical arrays of self-recording current meters, generally attached to lines stretched between a sub-surface buoy and a ground anchor.

Finally, brief mention should be made of attempts to measure total flow in some regions of the shelf seas by means of the magnetic induction effect; the flow of water in the presence of the earth's magnetic field induces a voltage at right-angles to the flow direction. Faraday was the first to suggest measurement of this voltage as a method of measuring total flow in a channel, and the idea has been investigated by Longuet-Higgins (1949), who calculated the distribution of electrical potential due to flow in a shallow channel of semi-elliptical cross section. He compared his calculations with measurements of ground potential at Lulworth due to tidal flow in the English Channel. Similar measurements were made by Brown and Woods (1971) at Aberystwyth on the eastern side of the St George's Channel in the southern Irish Sea. Other measurements use submarine telephone cables as sensors of the tidally induced potential. Bowden (1956) and Cartwright (1961) used the cable across the Strait of Dover to study tidal flow and mean flow through the Strait, and similar measurements were used by Hughes (1969) to measure tidal flow in the Irish Sea.

Synopsis of the tides

Tidal elevation

It is conventional to refer to the twice-daily tide due to the moon as the M_2 tide and similarly to that due to the sun as the S_2 tide. The progression of the tidal wave across the shelf is best depicted by charts of these tides which depict the co-tidal and co-range lines: co-tidal lines join places where high tide occurs simultaneously while co-range lines join places of equal tidal range. Whewell was the first to attempt to draw co-tidal lines spanning a sea area and in 1836 he published a tentative chart for the North Sea (shown, for example, in Proudman and Doodson, 1924). A controversial aspect of his chart was the suggestion that points of zero tidal range, known as amphidromic points, existed in the open sea, round which the tide revolved. It was not until the beginning of the twentieth century that observations confirmed the existence of amphidromic points and, in 1924, Proudman and Doodson published the definitive tidal chart for the dominant lunar semi-diurnal tides, the M_2 tides in the North Sea.

In constructing this chart, offshore measurements of tidal streams, assumed to be independent of depth, were used in the hydrodynamic equations to estimate the gradient of elevation at offshore points. When combined with elevation measurements at coastal stations, these gradients enabled the authors to draw elevation contours across the sea and hence map full co-range and co-tidal lines. Doodson and Corkan (1932a, b) used the same technique for the English and Irish Channels and published a complete chart of M_2 co-range and co-tidal lines for the seas surrounding the British Isles. Lennon (1961) subsequently extended these lines to the western edge of the continental shelf, and the charts are shown in Plate V.

The M_2 and S_2 components contribute most of the tidal energy in the British Seas and are each sinusoidal in profile. However, in a process which is analogous to the shoaling of wind waves discussed in the previous chapter, the tides change as they move from the deep ocean into the relatively shallow water of the shelf. The most important change from the point of view of the mechanics of seabed sand transport, which is discussed in a later chapter, is the generation of so-called harmonics. The term derives from the fact that the incoming tide with a period of about 12h generates a secondary component with half that period, i.e. with twice the frequency. Such changes in, for example, a musical scale are called musical harmonics and the same term is applied to waves. The change occurs because the wave form travels at a speed which is proportional to the water depth. The crest, however, is in relatively deep water and therefore travels faster than the trough so that the wave form becomes distorted. It is no longer a pure sinusoid but is composed of the primary wave and a secondary wave having twice the frequency of the primary. Thus the M_2 tide generates an M_4 harmonic which represents a lunar component with twice the frequency of the M_2 tide. This important effect produces an asymmetry in the tide, and is discussed further in Chapter 7.

The tidal motion of the continental shelf is not the result of the direct action of the gravitational forces of the moon and sun on the waters of the shelf seas. This was shown, for example, by G.I. Taylor, who, in his classic paper of 1918, estimated that the direct attraction of the moon on the waters of the Irish Sea contributes less than 7 per cent to the total energy dissipated there during a tidal cycle. The observed tides (Table 5.1) are, in fact, a response of the shelf seas to the tides generated in the wide expanse of the Atlantic Ocean. The Atlantic tides themselves are not as yet known in detail, but it seems certain that a counter-clockwise rotation of the M_2 tide about an amphidromic point south of Greenland, and a similar rotation about a second amphidromic near the Faroe Islands, north of Scotland result in a northward progression of the M_2 tide up the western edge of the continental shelf followed by a swing eastwards across the north of Scotland.

Table 5.1 Major tidal constituents

Location	Amplitude (cm)	
	M2	S2
Ocean tide (54°N)	8.53	3.98
Aberdeen	130.18	44.53
Holyhead	178.82	59.13
Devonport	169.13	60.26

The sequence of events which makes up the shelf seas' response to this Atlantic tide is best illustrated by the series of diagrams in Fig. 5.3(a–f). These diagrams show shaded contour plots of water elevation, derived from the co-tidal and co-range charts (Plate V) at 2h intervals over a complete semi-diurnal tidal cycle. The northward sweep of the Atlantic tide can be seen by following the region of high

Figure 5.3 Progress of the tidal wave around the British Seas.

elevation to the south of Ireland in Fig. 5.3(a) as it progresses up the western edge of the continental shelf in (b) and (c) and then turns eastwards round the top of Scotland in (d), (e), and (f) before flowing on into the North Sea. At the southern edge of the continental shelf, part of this northward-travelling Atlantic wave turns north-eastwards and passes almost perpendicularly over the shelf edge into the Celtic Sea, south of Ireland. The tidal wave then divides again around the south-west peninsula of England, part entering the English Channel and part entering the Bristol Channel and Irish Sea.

By the time the high water has reached the limit of the Irish Sea and English Channel the Atlantic tide has reached the entrance to the North Sea. In this broad sea the Coriolis force, which diverts movements to the right in the Northern Hemisphere, takes on a predominant role and causes the incoming high water to hug the east coast of Britain as it travels southwards. This southward-moving high water reaches the Strait of Dover about a tidal period after entering the North Sea, and thus joins the high water in the English Channel caused by the subsequent tide from the Atlantic. High water then turns eastwards and northwards, continuing to hug the southern and eastern boundaries of the North Sea to its right, but decreasing rapidly in amplitude and becoming negligible off the Norwegian coast.

Tidal currents

Tidal currents are more difficult to map, since both magnitude and direction of the flow at different times in the tidal cycle need to be displayed. Atlases of tidal streams, therefore, most commonly contain a sequence of charts showing the direction and strength of the currents at hourly intervals during a complete tidal cycle. Figure 5.4 shows contours of the peak spring tidal currents on the shelf.

The progress of the tidal wave northwards along the western shelf and around Scotland is predominantly that of a *progressive wave*, with peak flows corresponding to high and low water conditions. As such the tide is once again behaving in an analogous manner to the wind waves which were discussed in the previous chapter. Anyone who has stood in the breakers on a beach will know that the maximum onshore flow corresponds to the passage of the wave crest, whilst the maximum backwash occurs when the water is shallowest in the trough of the wave. Progressive waves have flow velocities which correspond to the wave height, and therefore the magnitude of the tidal flows increases with the local tidal range and hence with distance from the amphidromic point. Thus a comparison of Plate V and Fig. 5.4 confirms that the maximum flows in the North Sea occur along the coastline of England and Scotland, where the ranges are

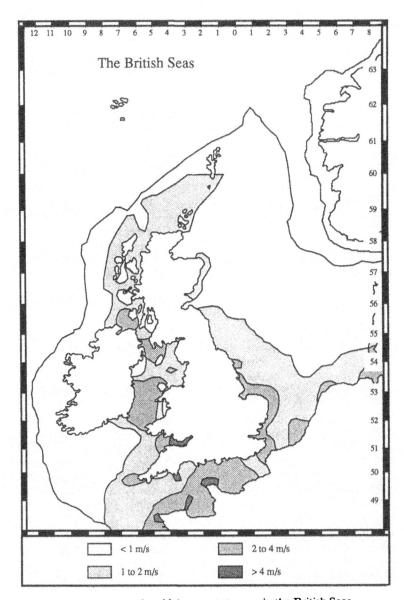

Figure 5.4 Maximum spring tidal current streams in the British Seas.

largest because the distance is greatest from the amphidromic points off the low countries and Norway. However, an entirely different process is operative in the English Channel and the Irish Sea. Tides there are dominated by a rocking motion with high water occurring alternately at the open and effectively closed ends of the basins, and a region in the centre where the tidal range is much smaller. This is characteristic of *standing wave* motion, and is confirmed by the existence of strong tidal streams towards the centres of the regions (at the standing wave nodes) even though the tidal ranges are relatively small. Again anyone who has watched an estuary fill from a pier or jetty will know that the maximum flood and ebb currents do not occur at high and low water, as would be the case with a progressive wave, but rather they occur at about mid-tide, for the estuary reflects much of the tidal energy and therefore represents a standing wave regime.

References

Bowden, K.F., 1956. The flow of water through the Straits of Dover related to wind and differences in sea level. *Phil. Trans. R. Soc. Lond., A*, 248, 517–551.

Brown, G.M. and W.G. Woods, 1971. Tidal influence on earth currents at a coastal station. *J. Atmos. Terr. Phys.*, 33, 289–293.

Cartwright, D.E., 1961. A study of currents in the Strait of Dover. *J. Inst. Navig.*, 14, 130–151.

Doodson, A.T. and R.H. Corkan, 1932a. The principle constituent of the tide in the English and Irish Channels. *Phil. Trans. R. Soc. Lond., A*, 231, 29–53.

Doodson, A.T. and R.H. Corkan, 1932b. New tidal charts for British waters. *Geogr. J.*, 79, 320–323.

Howarth, M.J. and D.T. Pugh, 1983. Observations of tides over the continental shelf of North West Europe. In: Johns B. (Ed.) *Physical Oceanography of Coastal and Shelf Seas*. Elsevier, Amsterdam, pp. 135–188.

Hughes, P., 1969. Submarine cable measurements of tidal currents in the Irish Sea. *Limnol. Oceanogr.*, 14, 269–278.

Huntley, D.A., 1979. Tides on the North West European continental shelf. In: Banner, F.T., Collins, M.B. and Massie, K.S. (Eds) *The North West European Shelf Seas: the Sea Bed and the Sea in Motion*, Vol. II *Physical and Chemical Oceanography, and Physical Resources*. Elsevier, Amsterdam, pp. 301–352.

Lennon, G.W., 1961. The deviation of the vertical at Bidston in response to the attraction of ocean tides. *Geophys. J. Roy. Astron. Soc.*, 6, 64–84.

Longuet-Higgins, M.S., 1949. The electrical and magnetic effects of tidal streams. *Mon. Not. Roy. Astron. Soc., Geophys. Supp.*, 5, 285–307.

The oceanography

Proudman, J. and A.T. Doodson, 1924. The principal constituent of the tides of the North Sea. *Phil. Trans. R. Soc. Lond., A*, 224, 185–219.

Pugh, D.T., 1987. *Tides, Surges and Mean Sea Level*. Wiley, Chichester.

Taylor, G.I., 1918. Tidal function in the Irish Sea. *Phil. Trans. R.Soc. Lond., A*, 220, 1–33.

76

Chapter six

The oceanographic regime

The oceanography of the British Seas includes the wave and tidal regimes detailed in earlier chapters, together with the chemical composition, physical conditions and biological activity of the water masses in the area. In the present chapter, the techniques involved in measuring the temperature and salinity of sea water are discussed, together with the results that these provide concerning the movement of water masses across the region.

Early measurements in the region were taken under the auspices of various oceanographical and meteorological cruises and during the 1930s the International Council for the Exploration of the Sea (ICES) nations began to co-ordinate sampling schemes from merchant ships on regular crossings in the area. ICES is a body which, since its inception in 1903, has monitored the fish stocks within the region, and now continues to do so within wider terms of reference. Modern research vessels have augmented these datasets, and a relatively complete synopsis of the distribution of temperature and salinity, and of the general circulation patterns in the British Seas can now be constructed.

The synopsis is presented in this chapter, but the subsidiary objectives are to provide the background information for more detailed work in later sections. In particular, it is the annual variations in water temperature and water stratification which control the movements of fish species, and to a certain extent therefore the operation of the fishing industry which is detailed in Chapter 10. Again it is the residual circulation patterns which control the large-scale dispersion of the pollutants which are detailed in Chapter 13.

Plate 6 Oceanographic measurements are made by deploying self-recording instrument packages from research ships.

Temperature

Heat budget of the world ocean

The world ocean has a surface area of some 360 million km², made up of a series of large interconnected basins and smaller continental seas. In order to understand the distribution of thermal energy in the British Seas it is therefore necessary to begin by considering the source of the heat, which most directly is the North Atlantic, and less directly is the world ocean as a whole. The thermal energy in the world ocean itself is very largely supplied from outside the earth and principally from the sun. There is a constant inflow of energy from the sun and a constant outflow of radiation from the earth into space. These considerations lead to the concept of a so-called steady-state system as far as the heat energy of the earth taken as a whole is concerned. This constancy of heat energy can be confirmed for the solid earth, that is the lithosphere, and for the atmosphere and it can be expected to hold for the oceans so that the supply of heat to the world ocean taken as a whole is balanced by an equally large loss of energy.

The largest thermal energy source is heat absorbed from solar and sky radiation. At the upper limit of the earth's atmosphere this produces, on average, an input of some $700 \text{gcal cm}^{-2} \text{day}^{-1}$. This is a large amount, because 1 gcal (pronounced gram calorie) is equivalent to the heat required to raise the temperature of 1g of water by 1°C. About 57 per cent of the energy is absorbed by the atmosphere so that $0.21 \text{gcal cm}^{-2} \text{min}^{-1}$ reaches the sea surface as shown in Fig. 6.1(a). Some 27 per cent is direct solar radiation and 16 per cent is diffuse radiation from the sky, known as sky light. The other sources of thermal energy for the ocean are much smaller and can be neglected. The heat obtained from the solid earth is, for example, on average $10 \times 10^{-5} \text{gcal cm}^{-2} \text{min}^{-1}$ and the disintegration of radioactive material in seawater generates about $4 \times 10^{-3} \text{ gcal cm}^{-2} \text{min}^{-1}$. The heating due to tidal energy (Chapter 5) being dissipated in frictional drag is only important in very shallow waters. For example, Taylor (1918) calculated a value of $0.002 \text{gcal cm}^{-2} \text{min}^{-1}$ for the Irish Sea, which if it could accumulate for a whole year would only lead to a temperature rise of 0.2°C.

Heat is lost from the world ocean by a combination of three processes. First, the ocean surface radiates energy back into the atmosphere and into space; second, convection currents transfer energy into the atmosphere immediately above the water surface; and third, the evaporation of sea water removes heat from the world ocean. The ocean thermal energy budget is thus written:

79

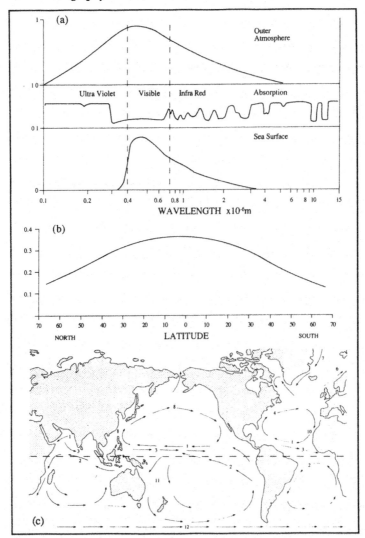

Figure 6.1 (a) Incident and absorbed solar radiation at various wavelengths for the upper atmosphere and the sea surface relative to the maximum, (b) the variation of the radiation input with latitude and (c) main residual ocean circulation patterns. 1, N. Equatorial Current; 2, S. Equatorial Current; 3, Equatorial Counter Current; 4, Gulf Stream Current; 5, North Atlantic Current; 6, Norwegian Current; 7, E. Greenland Current; 8, N. Pacific Current; 9, Labrador Current; 10, Canary Current; 11, E. Australia Current; 12, West Wind Drift.

$$Q_s - Q_b - Q_h - Q_e = 0$$

where Q_s is the input of solar and sky radiation, Q_b is the radiation from the sea surface, Q_h is the convection of heat from the sea surface, and Q_e is the heat loss due to evaporation.

This budget appears to hold for the world ocean as a whole, but the individual terms vary both spatially and temporally. The budget does not necessarily balance for any particular ocean or sea region and the imbalance leads to temperature and hence density gradients which result in ocean currents. Since, as shown in Fig. 6.1(b), Q_s varies with latitude, then equatorial waters are warmed, becoming less dense and rising to the surface to be replaced at depth by denser bottom waters moving towards the equator from higher latitudes. On a stationary earth the result would be simple convection cells, but the earth's rotation leads to the Coriolis force which tends to divert flows to the right in the northern hemisphere and to the left in the southern hemisphere resulting, in association with similar wind-driven currents, in the great ocean gyre systems which rotate in a clockwise direction in the North Atlantic and North Pacific and in an anticlockwise direction in the South Atlantic, South Pacific, and Indian Oceans. The oceanic circulation is further complicated by surface winds and by the ocean bathymetry. The resulting worldwide patterns are shown in Fig. 6.1(c). Here we are most interested in the circulation patterns in the North Atlantic, and it is clear that the Gulf Stream carries warm waters from tropical latitudes up the east coast of the United States and then, as the North Atlantic Drift, continues across to the western shelf boundaries of the British Seas.

The measurement of seawater temperature

Water temperature can be measured by a variety of techniques with varying accuracy and precision. Instruments commonly used include sensors based upon the expansion of a solid, liquid, or gas, or upon changes in the electrical properties of a sensor with temperature.

An example of the former class of instruments is the mercury thermometer which has long been used in oceanographic applications. This measures temperature as precisely as 0.01°C and is based on the thermal expansion of liquid mercury. Specially designed reversing thermometers are attached to water bottles and can be used to measure and record the *in situ* temperature at any depth. However, with the developments and improvements in electronics, thermistors have replaced mercury-filled thermometers as the preferred sensors in oceanographic temperature-measuring instruments. A thermistor is an electrical device, the resistance of which varies inversely with temperature. A small temperature change typically causes the resist-

The oceanography

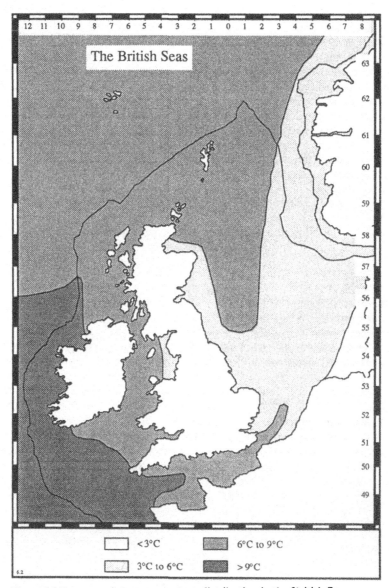

Figure 6.2 Average winter temperature distribution in the British Seas.
Source: Lee and Ramster (1981).

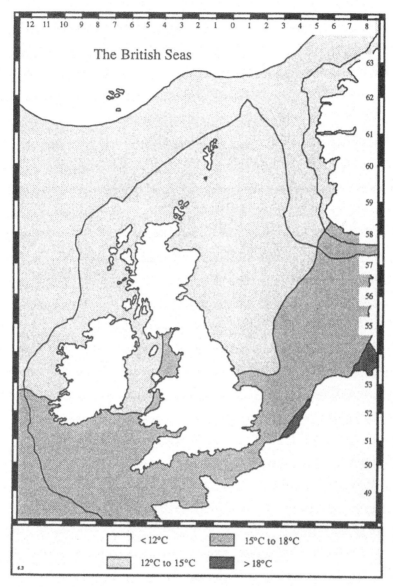

Figure 6.3 Average summer temperature distribution in the British Seas.
Source: Lee and Ramster (1981).

ance to vary by an order of magnitude. A thermistor is particularly appropriate for oceanographic use because the sensor can be made very small, it responds almost instantaneously to temperature variations, and it is capable of measuring temperature with a precision of 0.001°C. Thermistors are commonly integrated with salinometers into STD (salinity–temperature–depth) sensors.

Shelf temperatures in winter

Data have been collected using both mercury thermometers and thermistors throughout most of the British Seas region during the last 50 years. The resulting monthly mean surface seawater temperatures for February are shown in Fig. 6.2 and provide an indication of the coldest temperature field *on average* that occurs during the year.

The temperature ranges from more than 10°C in the south-western Celtic Sea to less than 3°C along the European coastlines. The coldest conditions are found in the shallow water off the coasts of Denmark and Germany, with the rest of the region being kept relatively warm by the North Atlantic water masses entering the Irish Sea from the south-west and the North Sea from the north and northwest. These derive from the North Atlantic Drift which is the continuation of the Gulf Stream shown in Fig. 6.1(c). During particularly cold winters (as in 1947, 1963 and 1976) ice fields form off the coast of the Netherlands, Germany and Denmark because these regions are closest to the cold continental land masses of Euro-Asia and farthest from the moderating influence of the North Atlantic.

Shelf temperature in summer

The monthly mean surface seawater temperatures for August are shown in Fig. 6.3 and provide an indication of the warmest temperature field *on average* during the year. The temperature ranges from more than 18°C in the European coastal areas to less than 12°C along the north-western shelf margin. The warmest conditions occur in the shallow water along the coasts of Denmark, Germany, and the Netherlands because they are closest to the land mass of Euro-Asia which, in summer, is considerably hotter than the water moving into the region from the west.

A striking feature of the distribution of temperature in both the winter and the summer in the British Seas, is the contrast between the east–west alignment of the isotherms to the west and south of the British Isles, and their north–south orientation in the North Sea basin. The reason for this change of alignment is, of course, the interplay between on the one hand the input of Atlantic water from the

84

west which has a relatively stable temperature of 10°C, and the extreme influence of the land mass conditions on the relatively shallow and enclosed basins of the North and Irish Seas. The Euro-Asian land mass fluctuates between sub-zero temperatures and more than 20°C during the annual cycle.

By following the lobes of water of a given temperature or temperature range one can obtain an idea of the position in winter as opposed to summer of the water moving into the area from the Atlantic. Clearly stronger incursions are made into the English Channel, Irish Sea, and North Sea in the winter than in the summer, but it must be noted that such changes are the result not only of water mass movement, but also of such factors as air–sea interactions and river run-off.

Salinity

The chemistry of the world ocean

The dissolving power of the water molecule leads to the presence of a great number of elements, although mainly in minute quantities, in any sample of seawater. Table 6.1 shows the major constituents of seawater.

The remaining elements are present in concentrations of less than 1 part per million (ppm). Every kilogram of seawater in the open ocean contains about 35g of ions. In oceanography this concentration is expressed as parts per thousand and is written 35‰, the ‰

Table 6.1 Major constituents of seawater

Ion	Symbol	% by weight
Cations		
Sodium	Na^+	30.62
Magnesium	Mg^{2+}	3.68
Calcium	Ca^{2+}	1.18
Potassium	K^+	1.10
Strontium	Sr^{2+}	0.02
Anions		
Chloride	Cl^-	55.07
Sulphate	SO_4^{2-}	7.72
Bicarbonate	HCO_3^-	0.40
Bromide	Br^-	0.19
Borate	$H_2BO_3^-$	0.01
Fluoride	F^-	0.01

sign being in accord with the per cent sign % which refers to parts per hundred. The measure of the concentration of dissolved ions is called salinity and the open ocean value can increase to as high as 40‰ in hot, dry regions where excess evaporation causes concentration of the ions, such as occurs in the Red and Dead Seas. Alternatively, freshwater inputs in cold, wet regions frequently dilute nearshore seawaters. Despite the range of salinities in seawater, the ratio of the elements given in Table 6.1 is remarkably constant and led Dittmar to conclude from his work on the famous *Challenger* expedition of 1872–6 that regardless of how salinity varies from place to place, the ratios between the amounts of the major ions in the water of the open oceans are constant. This has become known as the *Rule of Constant Proportions*. The table shows that common salt, NaCl, is the major constituent of seawater.

The measurement of salinity

The salinometer is now the most frequently used instrument for ocean salinity determinations. As will be discussed in some detail below, it is an instrument which measures water conductivity and temper- ature, from which the salinity can be computed. However, the salinity can be measured by a variety of techniques. The classical procedure requires a water sample which is titrated with silver nitrate. The result is compared to the precipitate from a 'standard seawater sample' to determine the salinity of the water. This method has a precision of approximately 0.01‰ or grams solute per kilogram of seawater. For convenient, quick estimates of surface salinity to within 1‰, a refractometer is often used. It is a relatively inexpensive piece of equipment, which may be sufficient for a pre-study of salinity distribution. It is based on the fact that the refractive index of water at 15°C varies from 1.3334 for a salinity of 0‰ to 1.3399 for 35‰, and thus the refraction pattern produced when a beam of light passes through a droplet of seawater will vary depending on the salinity of the seawater.

However, *in situ* measurements of electrical conductivity are presently preferred in ocean salinity surveys. They are quick and easy to make and yield approximately the same precision as the best titration method. As the salinity is not constant for a given conductivity but varies with the water temperature, it is essential to make simultaneous measurements of both the temperature and conductivity of the water sample.

The electrical conductivity of water is the inverse of its resistance and is either measured with an electrode system, where the conductivity cell is connected to form one arm of a Wheatstone bridge, or an inductive system, where the cell consists of two coils wound around an insulating core. An alternating current in the first coil induces a secondary current in the other coil through the core and the surrounding water. The second coil measures the induced current, which is in turn directly proportional to the electrical conductivity of the surrounding water.

Shelf salinities in winter

Once away from coastal influences the salinity of the world oceans varies only between the approximate limits of 32.5–37.5‰ so that analyses have to be conducted carefully in order to construct charts of salinity distribution. However, data have been collected over the last 50 years in the British Seas region and the typical distributions in the surface waters in winter are shown in Fig. 6.4. Once again the control influence is due to the North Atlantic Drift and, in general, areas with salinity greater than 35‰ contain Atlantic water, whilst those with a salinity of 31–32‰ show the influence of run-off from the European rivers.

Shelf salinities in summer

Fig. 6.5 gives the mean distribution of surface salinity in August and the most striking feature is the comparatively small area of the North Sea which is enclosed within the 35‰ isohaline. This confirms the pattern of the temperature data, suggesting that the influence of the Atlantic on conditions in the North Sea is less in summer months than in the winter. A similar retreat of the isohaline can be seen in the Irish Sea, the Celtic Sea, and the English Channel.

Residual circulations

Although the regular flow of water due to tidal currents in the British Seas is important for some short-term aspects of, for example, the movement of seabed sediments and the local flushing of coastal regions, it is the long-term or so-called residual water flows which control the large-scale dispersal of the type of pollutants which are discussed in Chapter 13. Attempts have therefore been made to construct annual mass balance budgets for the seas around the British Isles which take into account residual tidal circulations, the effect of large-scale currents entering the region across the shelf and from the Baltic, local river inputs, and estimates of precipitation and evaporation rates. The most complete is the following account which is based on a water mass budget for the North Sea constructed by Lee (1979).

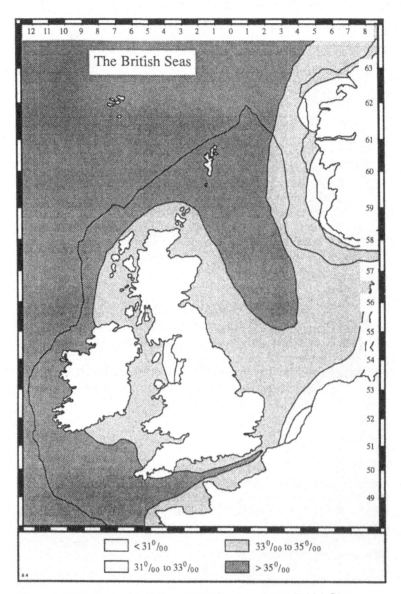

Figure 6.4 Distribution of surface salinity in winter in the British Seas.
Source: Lee and Ramster (1981).

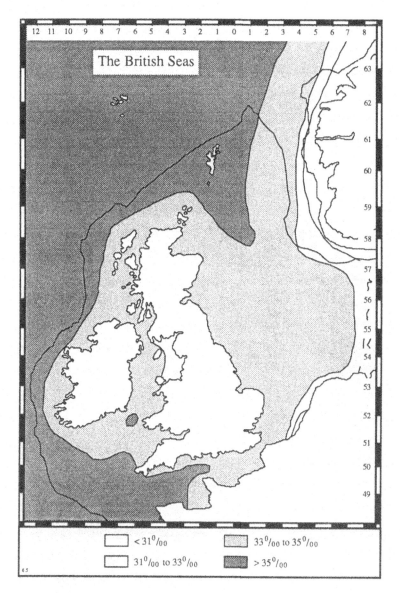

Figure 6.5 Distribution of surface salinity in summer in the British Seas.
Source: Lee and Ramster (1981).

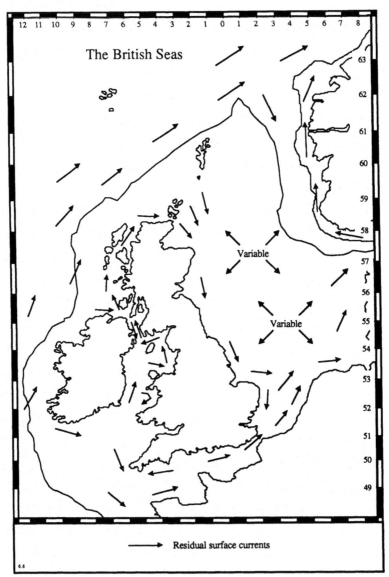

Figure 6.6 Residual near-surface circulations.

Source: Lee and Ramster (1981).

Water budget of the North Sea

Although shallow (Chapter 1), the relatively large surface area (575,000km^2) of the North Sea results in a total volume of some 54,000km^3. Atlantic water with a salinity of greater than 35‰ enters the North Sea through the Strait of Dover in the south and between Shetland and Norway in the north. The most recent and reliable estimate of the inflow through the Strait of Dover is due to Otto (1976) with a value of 3,400km^3 year^{-1}, equivalent to 6.3 per cent of the total volume. There are two estimates for the inflow from the north, the first of which has little observational basis and is due to Kalle (1949) with a value of 23,000km^3 year^{-1}. More recently, current measurements have suggested that the oceanic inflow in this region is mainly sub-surface and flowing along the western edge of the Norwegian Channel. Dooley (1974) estimates the northern inflow at 44,200km^3 year^{-1}, or 82 per cent of the total volume, 9,500km^3 of which enter between Orkney and Shetland. Additionally there are inputs from the Baltic (500km^3 year^{-1}, 9 per cent) from the river systems of the Meuse, Rhine, Weser and Elbe on the Continent and Thames, Humber, Forth–Tay and Moray on the UK coast (400km^{3}year^{-1}, 7 per cent) and from precipitation (400km^{3}year^{-1}, 7 per cent).

The major outflow from the North Sea is also along its northern margin, where waters from the Skagerrak and the northern region entrain some of the Atlantic waters to produce a flux of 63,000km^3 year^{-1} (Dooley, 1974). Some water returns to the Baltic (400km^3 year^{-1}, 7 per cent) and an equal amount is lost by evaporation. These produce an excess outflow of 14,900km^{3}year^{-1} which Lee attributes to a lack of knowledge of the actual size of some of the terms and suggests that the inflow and outflow between Shetland and Orkney should be increased and decreased by an amount equal to a half of the discrepancy, becoming 42,400km^3 year^{-1} (78 per cent) and 55,300km^3 year^{-1} (98 per cent) respectively. The turnover time, which is the ratio of the volume change per annum to the total volume, then has a value of very nearly one year.

Residual near-surface circulation

The non-tidal or residual current system shown in Fig. 6.6 is best thought of as indicating the most likely passage of an object floating submerged in the upper half of the water column for at least a year. Its rate of travel will be about 2–3km per day. The data show that the region divides into six sets of residual currents.

1.   The North Atlantic Drift is the warm continuation of the Gulf

Stream and flows, though with some variability, from the south-west in a generally north-easterly direction along and off the shelf edge. Three important spurs detach from the current and form:

2. The Celtic Current flowing in across the shelf and towards and into the western English Channel.
3. The Shetland Current flowing into the North Sea across the broad distance between the north of Scotland and the Shetland Isles.
4. The deep Norwegian Channel Current flowing in across the broad continental shelf edge and, at depth, south and eastwards around the coast of Norway. This current appears to predominate during the winter months.
5. The Shetland Current continues down the east coast of Great Britain and links with water flowing eastwards through the Strait to form the Southern Bight Drift which moves north and eastwards paralleling the coasts of Belgium, Holland, Germany, and Denmark before entering the Baltic Sea.
6. The Baltic Current flows eastwards through Skagerrak and then, hugging the coast of Norway, it turns northwards towards North Cape.

Currents in the English Channel, Irish Sea, the Malin Sea, and the Sea of the Minches are more complex, or perhaps simply more is known about them, but there is certainly a net northern drift through the western seas and around the northern coastline of Scotland, as will become apparent when the dispersion of radioactive waste material is described in a later chapter.

Oceanographic fronts

The British Seas have a vertical structure which varies substantially both from place to place and throughout the year. The differences occur because of the rate at which heat is transferred downwards due to internal mixing by wind and tides. When mixing is strong and the water depth is shallow then the heat is distributed evenly throughout the water column, but when mixing is weak or the water is deep then a warm surface layer forms above a cold bottom zone and the water column is said to be stratified. There is often an extremely sharp interface between these two layers called the thermocline across which temperatures may vary by as much as 8°C (Lee and Ramster, 1981). The chief factors determining whether or not an area is stratified are the water depth and the mean current speed and these have been

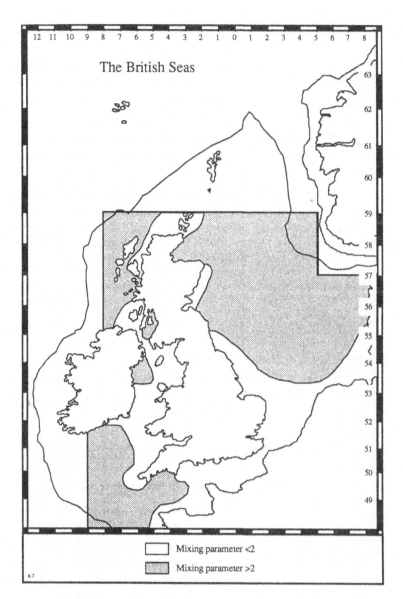

Figure 6.7 Distribution of the mixing parameter.
Source: Lee and Ramster (1981).

quantified by the use of the mixing parameter, M_p, where (Simpson *et al.*, 1977):

$$M_p = \log_{10} \frac{h}{u^3}$$

where h is the mean water depth (m) and u is the maximum tidal stream velocity at the surface (m s^{-1}). Values for M_p vary (Fig. 6.7) from greater than 4 in the northern North Sea and central Celtic Sea, indicating deep water, low current speed, and hence the presence of stratification, to less than 1 in North Channel, St George's Channel, the central English Channel, the southern North Sea, and certain coastal areas: locations where the shallow depth and high current speeds generate a fully mixed water column.

The transition from a stratified to a well-mixed area is often sharp, and gives rise to very high horizontal temperature gradients, local flow convergences, and often upwelling of unstable mixed but warm deep waters. These sharp transition zones are called *fronts*, being analogous to the atmospheric fronts shown on weather maps but the oceanographic fronts are more stable in position than their atmospheric counterparts. They can be associated with zones of high biological production and will later be seen to correlate well with the fishing grounds (Chapter 10). The fronts generally lie along the $M_p = 2$ lines as shown in Fig. 6.7.

References

Dooley, H.D., 1974. Hypothesis concerning the circulation of the northern North Sea. *J. Cons. Int. Explor. Mer.*, 36, 54–61.

Kalle, K., 1949. Die Natürlichen Ergenschaften der Gewässer. *Handb. Seefisch. Nordeuropas*, 1(2), 1–37.

Lee, A.J., 1979. North Sea: Physical Oceanography. In: Banner, F.T., Collins, M.B. and Massie, K.S. (Eds) *North-West European Shelf Seas: The Sea Bed and the Sea in Motion*, Vol. II. Elsevier, Amsterdam, pp. 467–493.

Lee, A.J. and J.W. Ramster, 1981. *Atlas of the Seas Around the British Isles*. MAFF, Lowestoft.

Otto, L., 1976. Problems in the application of resevoir theory to the North Sea. ICES CM 1976/C:18, 17 pp.

Simpson, J.H., D.G. Hughes and N.C.G. Morris, 1977. The relation of seasonal stratification to tidal mixing on the continental shelf. In: Angel, M.V. (Ed.) *A Voyage of Discovery: George Deacon 70th Anniversary Volume*. Pergamon, London, pp. 327–340.

Taylor, G.I., 1918. Tidal friction in the Irish Sea. *Phil. Trans. R. Soc. Lond., A*, 220, 1–33.

Chapter seven

Modern shelf sediments

Although, for very many centuries, mariners have been aware of local aspects of offshore seabed sediments in the British Seas, and of the menace created by the shifting of these sediments around tidal channels and sandbanks, it was not until the gentlemen geologists of the mid-nineteenth century that modern shelf sediments received any serious scientific attention. Then de la Beche (1851) and Lyell (1853) suggested that tidal currents could be responsible for the transport of sediment, and Reade (1888) suggested that such currents might move sediment far from its coastal site of origin, and that the passage of the sands should abrade the underlying seafloor. Dangeard (1925) and later Pratje (1950) progressed the debate by pointing out the good geographical correlation between the (by then well-known) distribution of tidal current speeds in the English Channel and the (then rather less well-known) distribution of seabed sediments in the same area. Later work has confirmed their suggestion that the boundary between mean peak spring, near-surface tidal currents (Chapter 5) of less than and greater than 100cm s^{-1} corresponds well with the change from sand to gravel-sized sediment. Additionally van Veen (1935) had used the echo-sounder to draw attention to the numerous sandbanks in the southern bight of the North Sea lying approximately parallel with the strongest tidal flows, and to the associated transverse sand waves which appeared to be moving northwards with the dominant tidal stream and to resemble many of the features of aeolian desert dunes.

However, even as recently as the decade after the Second World War, the existing seabed samples and the use of echo-sounders had revealed that there was a great complexity of sediment sizes and size distributions, and that the sand waves were but one of many types of active bedforms spread without apparent reason across the floors of the British Seas.

It was, however, the progress in underwater acoustics in the Sec-

Plate 7 Seabed sediments being recovered with a Shipek grab in the English Channel

ond World War, and specifically the development of the echo-sounder into the technique of side-scan sonar, that was to increase by many orders of magnitude the quality of the seabed data, and to provide answers to the confusion of complexity. The seemingly random variation in the nature of the sediments on the continental shelf which had so puzzled earlier workers was now recognized as a highly organized pattern of sand in transit, and it was seen that the shape and orientation of the continental shelf bedforms were clear markers to the routes of the great sand pathways around the British Seas.

The present chapter is based largely on the authoritative text by Stride and colleagues (1982) and details the techniques of seabed sediment sampling, and of side-scan sonar which have been used to identify these pathways, and then summarizes the present-day distribution of sediment on the UK Shelf. An understanding of this distribution is particularly important in the context of the resource chapters which appear later in this book. For example, certain types of commercial fisheries (Chapter 10) are based on species which spawn only on a gravel substrate. Alternatively the seabed mining industry (Chapter 11) is engaged in the exploration for and exploitation of specific seabed sediments. The present chapter will therefore draw extensively on preceding sections in order to provide a framework for the analysis of resource exploitation which is presented in Part II of this book.

Seabed sampling

The study of submarine sediments has been likened to mapping soil types in a mountain range from a helicopter in a thick fog. The analogy is apt and in either case a primary requirement is to obtain a sample of what is there and to determine the extent of particular types of surface. The techniques of seabed sampling are described in this section, and the modern mapping techniques using side-scan sonar are described in the following section. Reviews of sampling techniques are given by Bouma (1969) and the following summary is based on Shepard (1973). Sampling techniques can be divided into three classes: grabs, corers, and dredges; each will be described separately.

Grab samplers

The first seabed sediment samples were procured by arming the cup at the bottom of the sounding lead (Chapter 2) with tallow. This generally brings back a trace of the bottom sediment after each

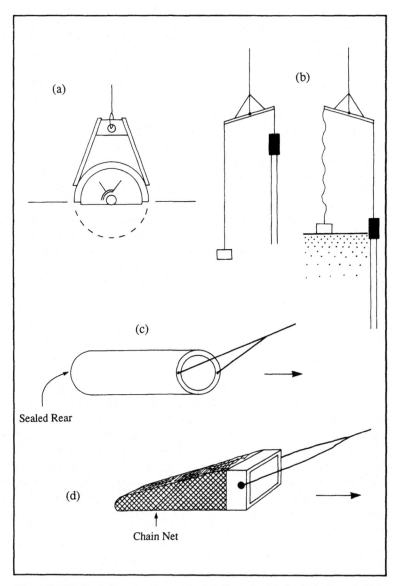

Figure 7.1 Methods for sampling seabed sediments: (a) the Shipek grab, (b) cantilevered gravity corer, (c) pipe dredge, and (d) frame dredge.

sounding, and North Sea fishermen were reputed to be able to navigate their ships from this information alone in bad visibility. Near the beginning of the century a more effective method was devised consisting of steel jaws which, held open during descent, closed upon striking the bottom by means of a spring or other type of release. Developed versions of these grabs are still in use today and have provided the majority of bottom samples in the British Seas. They range in size from small snappers which bring back about an egg-cupfull of sample and can be deployed by hand, to huge jaws that hold up to a cubic metre of material. The Shipek grab shown in Fig. 7.1(a) forms a cylinder when closed, which tends to prevent the stirring and loss of sediment during recovery; other designs utilize canvas tops for the same purpose. Grabs are most effective on soft seabeds and will recover consolidated muds and sands, although the spring-loaded devices tend to bounce and then close above the bed, returning only sea water.

Coring devices

Cores of the seabed are in many ways superior to grabs because they permit layering and structuring of the sediment to be distinguished. Again, a range of devices is commercially available from simple gravity corers which descend into the seabed under their own weight, through the cantilevered corer in Fig. 7.1(b), which only free falls the last few metres of descent, to vibro-corers which are lowered to the seabed and then forced into the sediment using compressed air in much the same manner as a road drill. Some corers have separate liners which are extracted from the barrel after recovery, whilst others are simple cylinders from which the sediment is extracted with a powered piston. Core diameters range from a couple of centimetres for small hand-held devices to 30cm or more for heavy duty, winch operated systems. Shepard (1973) details operating procedures for coring devices and notes that a serious problem is the foreshortening of the sample which occurs either during seabed penetration or during recovery. This effect can be partially alleviated by using internal pistons, but can still lead to underestimation of the sediment thickness. A development of the general coring principle is the box corer which, as its name suggests, is designed to recover a wide sample so that internal structures and bedding planes within the sediment can more easily be investigated.

Dredging

Most samples from solid rock outcrops on the seafloor have been ob-

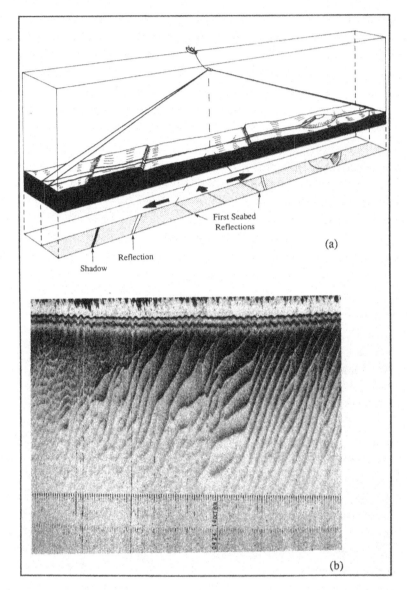

Figure 7.2 (a) Side-scan sonar technique and (b) typical output, showing a complex sand wave field in the southern North Sea.

Source: Photograph provided by N. Kenyon.

tained with either the pipe dredge shown in Fig. 7.1(c), or the frame dredge in Fig. 7.1(d). The pipe dredge is a brutish device consisting typically of a 2m length of steel pipe about 50cm in diameter, which is closed at one end. The open end is attached to a chain loop, which is in turn attached to a length of heavy anchor chain. The pipe is lowered to the seabed and then additional cable is let out as the pipe is towed along the seafloor, with the anchor chain serving to keep the dredge down. Samples knocked off the rock may then remain in the pipe during recovery. Rather subtler, the frame dredge consists of a rectangular steel opening which is also towed by heavy chain along the seabed. Samples entering the opening are retained in a chainmail basket. This device is rather better at keeping its sample during recovery than the pipe dredge, although finer particles are washed out of the basket.

Side scan sonar

Chapter 2 reviewed the development of echo sounding, whereby an acoustic pulse is emitted from the surface and the time interval which elapses until receipt of the returning echo from the seabed is measured, to compute the depth of water. The velocity of propagation of sound in seawater is about 1,460m s^{-1}. During the First and Second World Wars the technique was developed through the use of asdic (representing the initials of the Allied Submarine Devices Investigation Committee) which in the United States became known as sonar (SOund Navigation and Ranging). Side-scan sonar systems direct a series of lobes downwards and sideways from the ship or from a towed fish (Fig. 7.2(a)), and feed the returns to a stylus moving across sensitized paper. The pulse is emitted as the stylus leaves one side of the paper and since the stylus speed is kept constant the returning echo burns a darker patch into the paper at a distance from the leading edge which is proportional to the time elapsed for the echo to return and thus to the distance across the seabed from the ship's track. The paper is moved forward with each sweep of the stylus, and the system thus progressively reveals a map of the seabed features in proportion to the amount of acoustic energy which is reflected, as shown in Fig. 7.2(b). The results were stunning; wide swathes of the seabed could be surveyed relatively quickly and the variety of sediment types and bedforms revealed and mapped. It was the location and the orientation of the bedforms which was to be used by Stride's group to plot the direction of seabed sand transport in the British Seas, as described below.

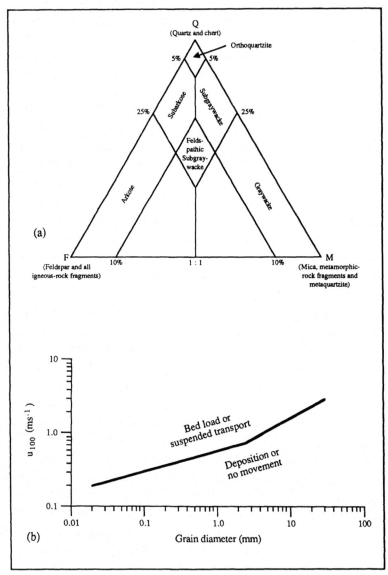

Figure 7.3 (a) Nomenclature for marine sediments based upon percentage composition, and (b) threshold velocities.

Source: After Shepard (1973).

Shelf sediments

Examination of sediments on the seabed reveals that they are composed of a variety of solid materials that can be grouped into five categories: terrigenous, biological, cosmic, submarine volcanic, and chemical. The number and kinds of each category in a particular sample depend upon the location of the sample on the shelf. Terrigenous and biogenous components are the most abundant in the British Seas.

Terrigenous sediment

Terrigenous sediments are produced by the weathering and erosion of rocks. These rock remnants are transported to the ocean by the action of rivers, waves, winds, and ice. Although some material is derived locally, such as the erosion of the Haig Fras granite in the Bristol Channel (Chapter 3), the majority of the modern shelf sediments are or have been eroded from British or continental land masses by the processes of chemical and physical weathering. During weathering each of the rock-forming minerals behaves somewhat differently. Calcite dissolves almost totally. The ferromagnesian minerals decompose and yield quartz, clays and dissolved material. Mica and feldspars behave similarly, except that mica reacts slowly and the feldspars slower still. In general the rock is reduced to:

(1) Quartz particles in the range 62μm to 1mm, which is called sand, and
(2) lithic gravel particles in the range 1–20mm.

Among the terrigenous sands the mineral quartz is the most abundant but large percentages of feldspars and ferromagnesian minerals including hornblende and pyroxene are also common. In addition mica, tourmaline, zircon, and garnet may be abundant. Many terrigenous sands have a large percentage of rock fragments. Depending upon the percentages of these constituents in a sand, a sediment may be classified as an *orthoquartzite* if predominantly quartz; an *arkose* if predominantly feldspar and sand-sized igneous rock fragments; or a *graywacke* if predominantly mica and metamorphic rocks. This classification is summarized by the triangular diagram shown in Fig. 7.3(a).

The distribution of sediment grain sizes in the British Seas has been determined from bottom sampling and is shown in Fig. 7.4 which confirms the correlation between sediment size and mean peak tidal current speed anticipated earlier. Compare Fig. 7.4 with the tidal currents shown in Fig. 5.5. This is because there is a well-

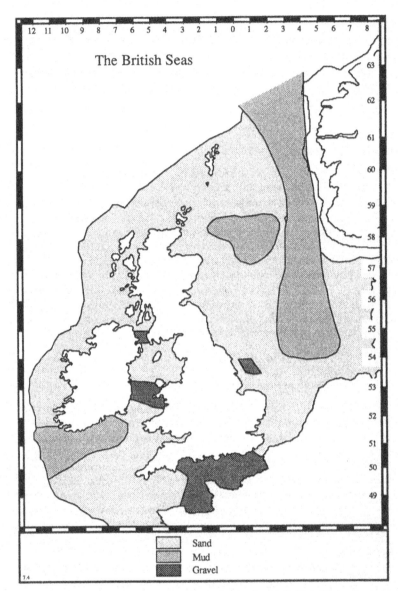

Figure 7.4 Distribution of sediment by grain size in the British Seas.

Source: After Stride (1982) and Allen (1970) and British Geological Survey 1:1,000,000 sheets.

proven relationship between the diameter of the sedimentary particles and the threshold flow speed at which they begin to move, as shown in Fig. 7.3(b). The anomalies are now understood in terms of (a) the local deeps and shoals of Pleistocene materials which are associated with local areas of anomalously fine or coarse grades of modern material respectively, and (b) sand that remains to be removed from a gravel floor. The observed boundary between sand and gravel does lie at about the position of the $100 cm\ s^{-1}$ line of near-surface mean spring tidal current strength. The conclusion is that the distribution of sand and gravel across the floors of the British Seas is approaching an equilibrium with the dominant tidal currents.

Biogenous sediment

The biogenous sands consist mostly of skeletal material of calcareous ($CaCO_3$)-secreting organisms, principally molluscan shells or shell fragments, including foraminifera tests, ostracods, bryozoans, and echinoids. In tropical areas, large quantities of coral and algal fragments are found in the sands. Siliceous organisms are usually smaller than sand size, but the sands may contain some of the larger diatoms and radiolarians. The calcium carbonate content of the sediment in the British Seas shows considerable variation (Wilson, 1982). The shell gravels of the western English Channel, for example, commonly have values of 60–80 per cent carbonate (Fig. 7.5). The sands forming the sand patches and the rippled sands on the continental shelf west of Scotland can contain 25–30 per cent carbonate while the coarser shell gravels between Orkney and Shetland commonly contain 90 per cent carbonate. In contrast the carbonate content of the active sandbanks of the southern North Sea is rarely more than 20 per cent and is generally about 5 per cent or less. Comparison of Fig. 7.5 with the sea temperatures (Figs 6.2 and 6.3) seems to suggest that, as might be expected, carbonate production and therefore sand carbonate content is related to the warmer waters to the south and west of the region, but this simple picture is undoubtedly complicated by the mobility of the seabed and the local habitat.

Shelf transport processes

The physics of the processes by which sands and gravels are transported by water in rivers and in the sea, and by wind in deserts has been investigated by engineers, geologists, geomorphologists, and sedimentologists. Although details are still to be clarified, some general principles have been derived and these can be applied to the

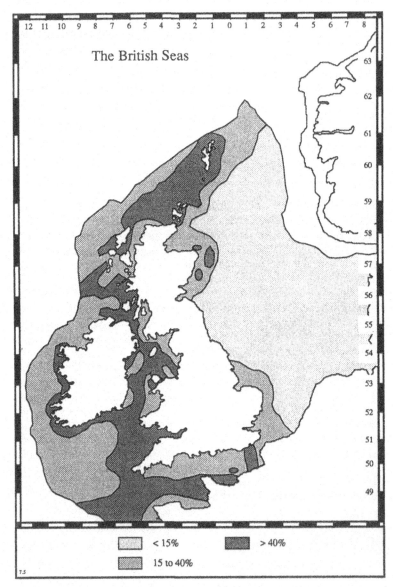

Figure 7.5 An impression of the abundance of biogenous debris in the uppermost sediments of the British Seas.

Source: Wilson (1982).

seabed sediments of the British Seas. It is apparent that the tidal flow does not actually transport the sediment until a certain threshold flow speed is exceeded and that, for flow measurements at a height of 100cm above the bed (u_{100}) the threshold speed (u_{100cr}) increases with the mean sediment diameter from about 20cm s^{-1} for a 0.1mm fine sand through about 60cm s^{-1} for a 1mm coarse sand to about 150cm s^{-1} for a 10mm gravel as was shown in Fig. 7.3(b). Once movement has been initiated, however, the rate of sediment transport, j, which is defined as the dry weight of sediment crossing a 1cm width of the bed in 1s is given by:

$$j = k(u_{1002} - u_{100cr2})\, u_{100}$$

There is, therefore, an approximately cubic relationship between the flow speed and the mass transport rate. Hardisty (1983) derived this relationship and tested it with sediment traps at 50°00'N, 7°01'W beneath 108m of water in the central Celtic Sea (Hardisty and Hamilton, 1984). The experiment showed that, for medium sand with a u_{100cr} of 27cm s^{-1}, tidal transport rates varied between 0.41 and 1.67mg cm^{-1} s^{-1} and that the calibration coefficient, k, was about 10^{-6}g cm^{-4} s^{-2}.

The movement of sediment on an initially flat sand bed results in accumulations of sediment into regularly spaced features known as bedforms in response to small and larger scale turbulent vortices within the flow. The type of bedform which develops depends upon the flow speed and ranges from small-scale ripples to large sand waves and dunes. The hierarchy of bedforms is listed in Table 7.1, and is illustrated for the waters around the UK by the block diagram in Fig. 7.6. Each will now be described in more detail.

Table 7.1 Hierarchy of tidal bedforms

name	scale	flow speeds (cm s^{-1})	flow type
Ripple	cms	> 20	Transverse
Sand wave	10–100 cm	> 60	Transverse
Parabolic sand wave	100 + cm	> 80	Transitional
Ribbons	10 + cm	>100	Longitudinal

Ripples

Although not detectable by side-scan sonar, sand ripples are almost ubiquitous on the seabed. They are distinguished from the larger scale sand waves by having (a) a wavelength less than 1,600 times and usually greater than 600 times the grain diameter of the sand and (b)

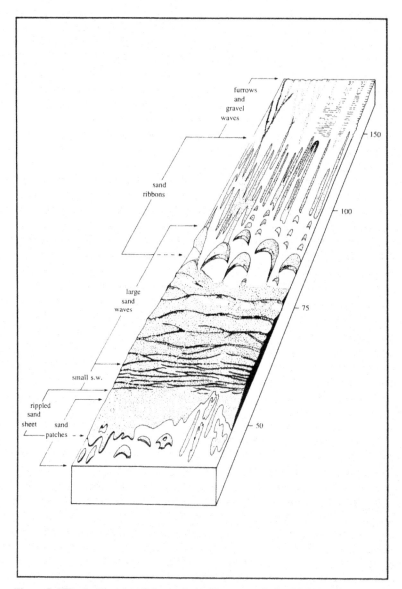

Figure 7.6 Block diagram of the main bedforms made by tidal currents on the continental shelf, with the corresponding mean spring peak near-surface tidal currents in cm s^{-1}.

Source: After Belderson, *et al.* (1982).

a height not greater than about 300 times the grain diameter of the sand. Wavelengths are thus less than about 60cm and are usually between 5 and 12 times the sand ripple height. They occur where flows are in excess of threshold, provided that the mean sediment grain size is not greater than about 0.7mm (Southard and Boguchwal, 1973), and are often superimposed on other bedforms.

Sand waves (dunes)

Sand wave is the term preferred by Belderson *et al.* (1982) for the larger, transverse bedforms which occur on the shelf. They typically range from 10 to 1,000m in wavelength and 1 to 15m in wave height. They are generated by tidal flows in excess of about 60cm s^{-1}, and small sand waves can form on the slopes of larger features. Sand waves are also formed by tidal lee waves at the continental shelf break and, on La Chappelle bank in the south-western Celtic Sea, these have wavelengths of about 1,000m and are up to 7m high (Stride, 1963).

Sand ribbons

The sand ribbons of the continental shelf vary greatly in size and can be as much as 15km in length. They range in thickness from a few grains to about 100 cm. Their edges are often sharply truncated, implying the existence of counter-rotating longitudinal vortices in the flow.

Shelf transport paths

In addition to the correlation of particular types of bedforms with the tidal current speed which was discussed earlier, the orientation of the bedforms is used to define the direction of net seabed sand transport. The transverse features such as ripples but more particularly the sand waves, have steep, lee avalanche faces in the down-path direction, whilst the sand ribbons lie parallel to the transport paths. This evidence was used in a compilation of side-scan records which was completed by Stride's group in the early 1980s at the Institute of Oceanographic Sciences of all of the available data on the direction of net sediment transport around the British Isles and is summarized in Fig. 7.7. Johnson *et al.* (1982) noted that such a compilation chart could be produced because there was such a striking measure of agreement about the main features and because of the encouraging agreement between the results and the directions and strengths of

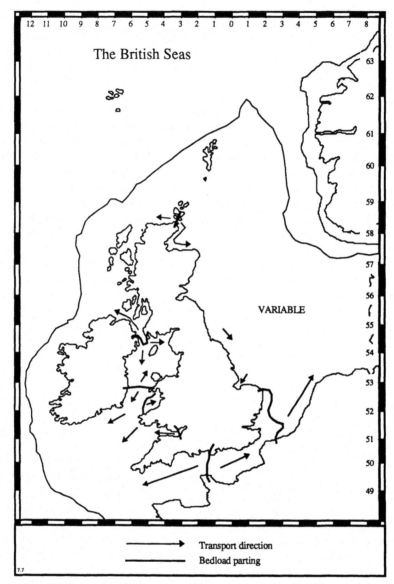

Figure 7.7 The net sand transport paths on the continental shelf around the British Isles.

Source: Johnson *et al.* (1982).

the local tidal flows (Chapter 5).

In some regions the net sand transport occurs on paths that extend parallel with the coast, whilst in others it may be at right-angles or some intermediate direction depending on the tidal current regime. Individual sand transport paths on the UK shelf are up to 550km (300 nautical miles) long and up to 170km (90 nautical miles) wide. The main paths originate at six major bedload partings and each of these will now be dealt with in some detail. In anticlockwise direction these are:

(a) Pentland Firth bedload parting

This, the smallest of the bedload partings, extends for about 20km in a roughly north–south direction from the northernmost promontory of Argyll in Scotland to the Shetland Islands. Sediment moves to the west along the coast towards the Sea of the Minches (Chapter 2) and towards the open shelf. Alternatively, sediment moves to the east and south into the Moray Firth or farther offshore, to join a south streaming path in the western north North Sea.

(b) North Channel bedload parting

This is also a relatively short bedload parting, lying between the west coast of Northern Ireland and Scotland. However, the line of the parting is far from direct, being located to the north in the shallow, coastal extremities but to the south in the deeper, central channel. Sediment moves in a north-westerly direction to the north of the parting into the deep waters of the Malin Sea, whilst to the south sediment appears to move into the Irish Sea or eastwards along the coast and into the extensive sandflat complex of the Solway Firth.

(c) St George's Channel bedload parting

This parting is about 140km long and runs first in a west–east direction from the Irish coast towards Wales before turning abruptly south-west to terminate off Milford Haven. To the north sediment is carried into the Irish Sea or into coastal waters, while to the south sediment is carried out into the Celtic Sea.

(d) Bristol Channel bedload parting

Another short parting has been identified in the upper reaches of the Bristol Channel along a line which, coincidentally, lies close to that being considered for a tidal barrage (Chapter 12). Upstream, sediment is moving inexorably into and filling the Severn Estuary whilst to the west it is moved offshore into the north-east to south-west system of the Celtic Sea.

(e) English Channel bedload parting

This is perhaps the best known bedload parting and runs for some 140km between the Isle of Wight and northern France. To the west, sediment is moved into the long and wide pathway which culminates many hundreds of kilometres further out in the heads of the canyons on the shelf edge. To the east, sediment is swept towards the coast of France and into the Strait of Dover.

(f) Southern Bight bedload parting

This, at some 400km, is the longest and most complex parting on the shelf and exhibits a sinuous path between the coast of England in Norfolk and France near Calais. Again the parting runs close to each coast in shallow water, but further offshore it turns and eventually forms a broad loop amongst the sandbanks off the Thames estuary. To the west, sediment is carried but a short way to a convergence in the Strait of Dover, but to the east the sediment is carried along the coast of the low countries and England into deeper, less well-known areas.

Tidal control on transport paths

It is apparent that the tidal current speed controls the size of the sediment, and that it controls the rate of sediment transport along the transport paths and the type of bedform which is developed across the shelf, and it therefore seems reasonable to investigate if it finally controls the direction of these paths. Initially the answer would appear to be in the negative for there is little evidence of, for example, the location of the bedload parting lines, in either the tidal streams or residual currents which were described in Chapters 5 and 6. However, upon further consideration this is not surprising, and a more plausible explanation based upon these tidal streams can be developed. First, it was shown in an earlier section that the sediment transport rate is roughly proportional to the cube of the current speed. Again, Chapter 5 demonstrated that the tidal streams, at least in very deep water, could be correlated with the lunar and solar gravitational components. However, these components generate a regular, symmetrical sinusoidal tidal flow, so that the sediment transport rate would itself be cubed but still regular, symmetrical, and sinusoidal. The result would be equal amounts of transport by both the flood and the ebb currents, which would obviously cancel one another and could not lead to net transport or to the development of a transport path.

However, the propagation of the tide across the shelf is not a

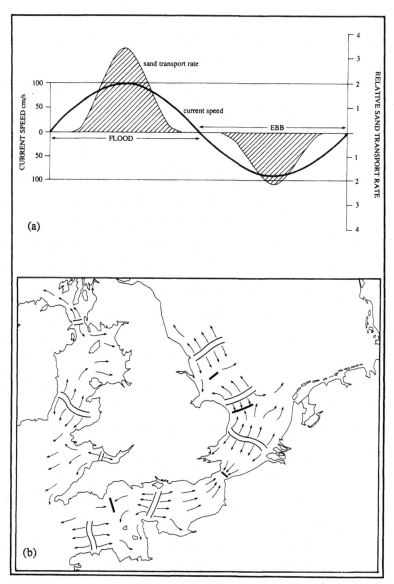

Figure 7.8 (a) Summation of the M_2 and M_4 tidal components to generate flow asymmetry and (b) numerical simulation of maximum flow due to M_2 and M_4 constituents.

Source: After Pingree and Griffiths (1979).

simple process and one of the most significant changes is the generation of the so-called tidal harmonics which were discussed in Chapter 5. In particular, we saw that the semi-diurnal M_2 with a period of just over 12h generates an M_4 harmonic with half that period. The addition of these two symmetrical sine waves does not, however, produce symmetry, but instead produces a flood tide which is of shorter duration but of higher peak flow speed than a longer duration, but less swift ebb as shown in Fig. 7.8(a). The sediment transport rate which is produced by cubing these flow speeds is now also no longer symmetrical but is dominated by the flood tide, producing net transport and opening the possibility of a tidal transport path. This idea was investigated by Pingree and Griffiths (1979) who constructed a computer model of the resultant of the sum of the M_2 and M_4 tidal constituents, and their predictions are shown in Fig. 7.8(b). These correspond remarkably well with the transport paths shown earlier. The North Channel, St George's Channel, Bristol Channel, English Channel, and even the complicated Southern Bight bedload partings are clearly depicted in the tidal chart, as are the directions of most of the transport paths emanating from these partings.

It appears then that the apparent complexity of sediment and bedform distribution across the bed of the British Seas can be elegantly explained in terms of the tidal flows, and particularly in terms of the flow asymmetries generated by the lunar tide and its harmonic.

References

Allen, J.R.L., 1970. *Physical Processes of Sedimentation*. Unwin University Books, London, 248 pp.

Beche, H.T. de la, 1851. *The Geological Observer*. Longman, Brown, Green & Longmans, London.

Belderson, R.H., M.A.Johnson and N.H.Kenyon, 1982. Bedforms. In: Stride, A.H. (Ed.) *Offshore Tidal Sands*. Chapman and Hall, London.

Bouma, A.H., 1969. *Methods for the Study of Sedimentary Structures*. Wiley-Interscience, New York, 458 pp.

Dangeard, L., 1925. Observations de geologie sous-marine et d'oceanographie relatives a la Manche. *Annales de l'Institut Oceanographique Paris*, 6, 1–295.

Hardisty, J., 1983. An assessment and calibration of formulations for Bagnold's bedload equation. *J. Sediment. Petrol.*, 53, 1007–1010.

Hardisty, J. and D. Hamilton, 1984. Measurements of sediment transport on the seabed southwest of England. *Geo-Marine Letters*, 4, 19–23.

Johnson, M.A., N.H.Kenyon, R.H.Belderson and A.H.Stride, 1982. Sand transport. In: Stride, A.H. (Ed.) *Offshore Tidal Sands*. Chapman and Hall, London.

Lyell, C., 1853. *Principles of Geology*. John Murray, London.

Pingree, R.D. and D.K.Griffiths, 1979. Sand transport paths around the British Isles resulting from M_2 and M_4 tidal interactions. *J. Mar. Biol. Assoc. UK*, 59, 497–513.

Pratje, O., 1950. Die Bodenbedeckung des Englischen Kanals und die maximalen Gezeitenstromgeschwindigkeiten. *Dtsch. Hydrogr. Z.*, 3, 201–205.

Reade, T.M., 1888. Tidal action as an agent of geological change. *Philos. Mag.*, 25, 338–343.

Shepard, F.P., 1973. *Submarine Geology*. Harper & Row, New York, 517 pp.

Southard, J.B. and L.A. Boguchwal, 1973. Flume experiments on the transition from ripples to lower flat bed with increasing sand size. *J. Sediment. Petrol.*, 43, 1114–1121.

Stride, A.H., 1963. North east trending ridges of the Celtic Sea. *Proc. Ussher Soc.*, 1, 62–63.

Stride, A.H. (Ed.) 1982. *Offshore Tidal Sands*. Chapman and Hall, London.

Veen, J.van, 1935. Sandwaves in the southern North Sea. *Hydror. Rev.*, 12, 21–29.

Wilson, J.B., 1982. Shelly faunas associated with temperate offshore tidal deposits. In: Stride, A.H. (Ed.) *Offshore Tidal Sands*. Chapman and Hall, London.

Part II

The resources

Chapter eight

Trade and shipping

The volume of cargo moved by sea increases each year. Statistical data are not at hand to show the development of seaborne cargo traffic since, say, the beginning of the industrial revolution, but there can be no doubt that the increase has been tremendous. Even in the shorter perspective of the present century, the evidence for a continued increase in merchant shipping traffic is impressive as shown by Fig. 8.1(a). The absolute perspective depends, of course, on the measure of transport load, and if the total exchange of commodities and services between nations is considered, instead of the simple gross tonnage, then seaborne cargo movements account for a declining share. This is for two reasons, first an increasing share of the commodity flow has been transported across land boundaries by road, rail, pipeline, and aeroplane, and second the service sector accounts for an ever-increasing share of the total exchange, particularly between the developed countries of the western world.

Nevertheless, Fig. 8.1(b) shows that, until the Second World War, the UK was the leading foreign trade nation in the world, and although this position has since been usurped by the United States and West Germany, the British Seas continue to harbour the most densely located collection of trading ports, and are therefore still the busiest sea lanes for both coasting and ocean traffic.

This chapter begins with a general review of world shipping and then details some of the workings of the major ports in the British Seas, for their economic success necessarily controls the seaborne traffic density. This background will be used to identify and to estimate the quantities of particular types of seaborne traffic. The chapter concludes with some comments upon the distribution of submarine telephone and data cables across the shelf, so that the distribution of both shipborne and cableborne trade in the British Seas can be fully appreciated.

Plate 8 Shipping operates between the major ports within the British Seas and also connects with overseas markets around the world

World shipping

At the beginning of the present century there were about 23 million gross tons of coal-burning steamer ships in the world fleet, and there remained a further 6.5 million tons of sailing ships. These sailing ships were far less efficient than the steamers, and although they still accounted for a significant part of all sea trade in 1900 they all but the smallest disappeared in the early years of the present century. The first diesel-engined ocean-going ship, the *Selandia*, was built in 1912. Very few steamships with reciprocating engines are now being built, but steam turbines propel a rapidly expanding tonnage and the total tonnage of steamships has been slowly increasing.

The world gross tonnage doubled in the 14 years from 1900 to the First World War and continued to grow in the inter-war years with the exception of 1924 and the depression years of the early 1930s. During the Second World War almost 7,000 ships totalling almost 34,000,000 gross tons were lost, but new buildings more than compensated for these decreases and after the war the fleet resumed a rapid increase: in 1961 there were 130 million gross tons of steamers and motor ships.

In recent years, vessels in the largest tonnage categories have made up an increasingly greater part of the world fleet. While in 1939 ships of less than 6,000 gross tons accounted for 57 per cent of the total tonnage, the corresponding 1961 figure was 23 per cent, with the 6,000–10,000 gross tons category being in the majority. Among the larger ships, the growing importance of large tankers is very obvious, particularly after the closure of the Suez Canal and the diversion of the world oil supplies around South Africa. In 1939 most tankers fell into the 6,000–10,000 gross tons category, whilst by 1961 the tanker tonnage consisted almost exclusively of vessels over 10,000 tons, and such ships of 100,000 tons or more were becoming common by the late 1960s.

The most up-to-date information on world shipping is to be found in annual publications such as Lloyd's Register (1988). These show that the total gross tonnage in the world freight fleet has risen steadily during the last quarter of a century but is presently relatively stable at just over 400 million tons as shown in Table 8.1. However, the tanker fleet, which continued to grow through the 1960s and 1970s, has been reduced by almost 30 per cent in the last decade. It appears that there are currently about 20 million gross tonnage of vessels constructed each year with a similar or slightly higher figure being broken up annually.

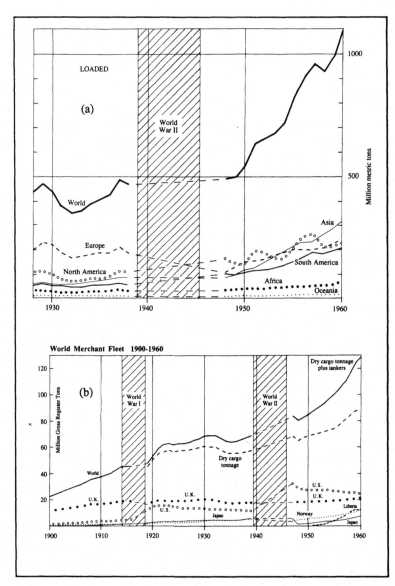

Figure 8.1 (a) International seaborne cargo 1928–60 and (b) gross shipping tonnage of the major nations.

Source: Alexandersson and Norstrom (1963).

Table 8.1 Gross tonnage in the world fleet 1965–88

Year	Total (million tons)	Tankers (million tons)
1965	171	60
1970	220	95
1975	310	130
1980	420	175
1985	420	150
1988	410	125

Major ports

The development and principal trades of the major ports in the ten countries bordering the British Seas will now be reviewed. The location of these ports is shown in Fig. 8.2 which is based on Lee and Ramster (1981), who also utilized Anon. (1971), Bird (1971), and Couper (1972). All of the ports shown on the diagram each dealt with more than 500,000 tonnes of cargo a year either as imports or exports in the 1970s and in most cases this criterion was reached for both imports and exports. The most important of the more recent port developments included the deep-water port at Cap d'Antifer near Le Havre, the new oil terminals at Sullom Voe and Bantry Bay, and the new iron ore terminal at Hunterston.

The following section separates the ports on the basis of geographical location, but they can also be classified in terms of their patterns of trade. The largest have a wide range of trades related to rather ill-defined national and international hinterlands, as well as to industrial development, e.g. Hamburg, the Weser Estuary, Ijmuiden-Amsterdam, Rotterdam, Antwerp, Le Havre, London, Humberside, Liverpool, Southampton, and Manchester, although the last named is declining rapidly as its ship canal falls into disuse. Smaller scale operations draw on a more restricted and better defined hinterland and include Leith, Glasgow, Dublin, Tyne-Tees, Bristol, Cardiff, Ghent, and Rouen. Finally in the remote, mainly rural areas are a number of ports such as Bergen, Aberdeen, and Cork with more clearly defined regional hinterlands and the development of, for example, oil-based activity. Behind this general classification lies a trend towards specialization in shipping and port development with port operations that have been developed around a few types of cargo now moving towards the development of separate oil terminals, bulk ore terminals, container terminals, and roll-on/roll-off facilities, often down river from the original docks. Information on these infrastructural and port developments is contained, for example, in Lloyd's Register (1988).

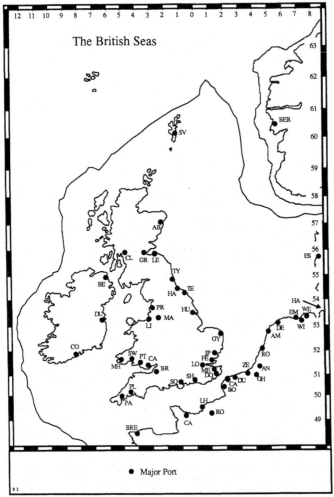

Figure 8.2 Major ports in the British Seas. British Isles: SV, Sullom Voe; AB, Aberdeen; GR Grangemouth; LE, Leith; TY, Tyne; HA, Hartlepool; TE, Tees; HU, Hull; GY, Great Yarmouth; IP, Ipswich; LO, London; ME, Medway; DO, Dover; SH, Shoreham; SO, Southampton; PL, Plymouth; PA, Par; BR, Bristol; CA, Cardiff; PT, Port Talbot; SW, Swansea; MH, Milford Haven; LI, Liverpool; MA, Manchester; PR, Preston; CL, Clyde; BE, Belfast; DU, Dublin; CO, Cork. Continental Europe: BE, Bergen; ES, Esjberg; HA, Hamburg; WE, Weser; WI, Wilhelmshaven; EM, Emden; DE, Delfzijl; AM, Amsterdam; RO, Rotterdam; AN, Antwerp; GH, Ghent; ZE, Zeebrugge; DU, Dunkerque; CA, Calais; BO, Boulogne; RO, Rouen; LH, Le Havre; CA, Caen; BRE, Brest.

124

Norway

Norway is much less prominent on a port map than in the world's shipping industry because her fleet operates largely between foreign customers, and, in fact, only 10 per cent of it calls at a Norwegian port in a normal year. The Norwegians developed their early fleet through whaling interests, and Sandefjord at the mouth of Oslofjord was the home port of the whaling factory ships. This usage has, of course, declined but Oslo and the other Oslofjord ports still account for the majority of modern Norwegian sea trade. This centre is, however, outside the area covered here and Bergen is the only major west Norwegian seaport. Its chief work is still concerned with the landing of fish and the exporting of fish products.

Denmark

Again the major Danish ports are to the east of the country, and although these clearly contribute to the Baltic Sea traffic which is described in the following section, they are outside the definition of the boundaries of the British Seas which is used here. However, the Danes controlled the Alborg Sound which was for centuries the safest passage from the North Sea to the Baltic ports in Sweden, East Germany, Poland, and Russia, and their maritime influence on the whole of the European Atlantic seaboard was therefore considerable. Copenhagen was for centuries the leading Danish and Baltic port but the construction of the Kiel canal made Hamburg a serious competitor for the transit trade.

Esbjerg is the only major port on the west coast of Denmark, and is a relatively new development. It was built in 1868 as a result of a decision in the Danish Rigsdag to create a west-coast port in response to the rapid development of Denmark as a major supplier of bacon, dairy products, and eggs to Britain. It became a railhead for these exports, and was shipping two-thirds of the country's bacon and more than one half of its butter and eggs in the years before the Second World War. After the war these percentages have been reduced as Britain's share of Denmark's agricultural exports decreased. Esbjerg is also the leading fish harbour in Denmark and is the home port of a large merchant fleet.

West Germany

Hamburg, the largest West German port, was founded in 825 but, throughout its early history, it was little more than the North Sea outport for Lübeck, the capital of the Hanseatic League, and involved largely in local trading. In the sixteenth century, however, trading horizons broadened and the port flourished to serve a massive and industrialized hinterland along the Elbe. Trade reached its zenith in the early years of the present century when it was the world's fourth largest port, and one of the earliest to operate a modern freeport system, free from the controls of local custom authorities. The re-shaping of Europe did, however, have a profound effect on Hamburg, largely because the Iron Curtain severed the ports hinterland only about 45km upstream from the city centre. Although Hamburg remains West Germany's premier port the European emphasis has shifted further south.

Bremen and the other ports on the Weser Estuary form the second largest port complex after Hamburg, and have developed transcontinental traffic with the United States. At Bremen's Bremerhaven a new ore-handling facility was completed in the mid 1960s. The old German naval base Wilhelmshaven, whose cargo tonnage was quite insignificant in the 1950s, has developed into one of the leading German ports by tonnage with the opening of the 28-inch crude petroleum pipeline to the Ruhr. Fulfilling a similar role, the port of Emden was developed in the 1890s as a transshipment point for the European canal network and particularly for the industrialized Ruhr. Coal, ore and grain have all formed major trade commodities through the port, and, more recently, new refinery facilities are making oil a major cargo so that Emden remains a bulk cargo port serving the industrialised Ruhr hinterland.

Holland

Both Holland and Belgium were, until recently, the home countries of large overseas empires, which were serviced by shipping from the ports clustered around the entrances to the Rhine. The decline of empire has not, however, been accompanied by a lessening of the importance of this location, and the trades thrive serving the highly industrialized and densely populated hinterland of lowland Europe. Rotterdam, which has become the world's leading port on a tonnage basis in the years following the Second World War, was founded in the Middle Ages but was overshadowed by neighbouring Delft. The greatest development came with the industrial revolution in Germany which started in about 1850 and turned the Rhine basin into the main manufacturing area of Europe. The new sea canal cut from the port in 1866–72 offered greatly improved access to the port and avoided the navigational problems of the Rhine delta. The development of the port as a trans-shipment site between river barges and sea-going ships was then inevitable and its vast complex nowadays

126

continues to service the iron and steel industries of the Ruhr, and also includes large petrochemical refineries with facilities for the attendant tankers and supertankers. This development has continued with the building of the massive Europoort complex which is connected to the German Rhineland through a 24-inch pipeline capable of pumping some 6.5 million tonnes of oil per year. Most recently there has been the development of the Maasulakte complex with bulk handling and container facilities to service the largest vessels.

Amsterdam, which was developed to serve the shallow-draft coasting trade in the Zuider Zee in the Middle Ages, also opened itself to the North Sea through the construction of a canal in 1876, and was for a long time the almost exclusive marketplace for products from the Dutch colonies. The development of the port also accelerated with the opening, in 1952, of the Amsterdam–Rhine canal and it has now become a major Rhine estuary port with modern facilities for dry bulk cargo and petroleum products. The associated port of Ijmuiden at the sea entrance of Amsterdam's canal has become the main location for the Dutch fishing industry and for iron and steel developments. Delfzijl is the largest port in northern Holland and specializes in the local and coasting trades.

Belgium

Until the increases in air travel over the last 30 years, the majority of the shipping calling at Antwerp consisted of cargo and passenger liners. The post-war development of petrochemicals has replaced this trade and the port remains Belgium's largest, dealing mainly with its national hinterland, although transshipment of barge traffic from further afield is important. It has the disadvantage of lock basin access, but the world's largest locks were opened here in 1989. The other Belgian ports of Ghent and Zeebrugge are sufficiently large to be included on Fig. 8.2, and are now concerned with the general cargo and ferry trades, and at Ghent with bulk cargoes of ores and fertilizers.

France

The French ports on the English Channel have a more favourable location than those on the Bay of Biscay because they serve an industrially advanced hinterland stretching from France into Germany and Belgium. The more easterly ports of Dunkerque, Calais and Boulogne serve the cross-Channel trade, the others also service ocean-going vessels, and between them they make a considerable contribution to the traffic of the British Seas.

The resources

The eastern ports all service a considerable passenger and freight ferry trade, although each has developed independently. Boulogne was perhaps the oldest cross-Channel route, being built from a rocky promontory by the Romans, but was surpassed in this trade in the medieval period by Calais, which was in fact an English base between 1347 and 1558. Boulogne is today France's leading fish landing port with fast freight services to the large urban centres, notably Paris. Calais is, in contrast, built on a dune coast, and the majority of its trade serves the shortest Channel crossing to Dover, with facilities for berthing at all states of the tide for the constant ferry shuttle service. Dunkerque is the chief city of maritime Flanders and was contested by France, England, and Spain before being finally attached to France in 1662. It has an excellent geographic location on the North Sea mouth of the English Channel and has long competed with Antwerp to export the material products, particularly steel, of the European low countries. Steel now forms the port's major export, and the trade was further enhanced in the 1960s by the construction of Europe's largest steel mill in the extended port area of Dunkerque. In addition, a large volume of car ferry traffic with Dover and Ramsgate, petroleum, and general cargoes are also handled.

The lower Seine ports of Rouen and Le Havre serve the French capital, which is over 100km further inland. The former developed on a Roman site some 106km from the sea, and evolved through textile and paper manufacturing. The decline of the port in the mid-nineteenth century was largely halted by the construction of training walls in the river to deepen the channel and now it is involved in grain import and export and the service of France's largest paper products industry. Most ships are engaged in short or medium length runs and are thus of moderate size. Le Havre has also been somewhat limited by the complicated hydrographic evolution of the lower Seine, but deep dredging has been used with effect. The construction of crude oil pipelines from the port to refineries in the Seine valley has combined with a trade in general cargo, notably foodstuff commodities, cotton, and copper to make Le Havre the second French port after Marseille. The completion of an outport at Antifer in the late 1970s to accommodate 500,000 tonne tankers was largely overtaken by the declining use of the largest crude carriers. The other major French channel port is Caen, which developed to export the iron ore deposits of Basse-Normandie. Local steelworks have also been developed and the port now rivals Dunkerque for steel exports. Further west, Brest is the only sizeable port in Brittany, and is largely a naval base with a relatively small commercial cargo turnover.

England

The UK was for centuries the mother country of a global empire and the world's leading nation in international trade. Although that position has now been taken by the United States, it is still the shipping bound to and from her ports which combines with continental container and bulk traffic to make the British Seas into the busiest shipping lanes in the world. The ports were originally developed at the crossing points of estuaries, where goods could be brought as far inland as possible, and shelter was provided. The increasing size of ships led, however, to a downstream development of most port complexes to provide deeper water access. The country's main port is London which was an important settlement in Roman times, and is still the capital city and financial centre although much of the industrial wealth emanates from further north. London's docks, in the east of the city, were cut from the soft alluvium of the banks of the river Thames to alleviate severe overcrowding of the river in the early nineteenth century, but much of the trade has now moved downstream to the Medway, to Sheerness, to the oil refineries on Canvey Island and at Shellhaven, and to the container port at Tilbury. The upriver dock systems have now been closed, or converted for pleasure craft.

The Merseyside complexes at Liverpool and, 55km inland, on the Manchester ship canal are England's second largest when taken in terms of combined tonnage. Liverpool developed as a trading port for the Atlantic, particularly with North America, whilst the ship canal brought sea-going vessels into the heart of textile country and also served the extensive petrochemical works at Ellesmere Port and Port Sunlight.

Coal mining, steel making, shipbuilding, and engineering have decided the pattern of ports on the north-east coast of England. Newcastle, the major Tyne port, developed to serve the coastal collier trade, but has more recently grown to provide facilities for iron ore, grain, lumber, cement and a long list of cargoes to serve its heavily industrialized hinterland. Middlesborough and the other Tees ports farther south specialize in steel and petrochemicals, and farther north, Hartlepool is the site of a once great shipbuilding industry now maintained with defence orders. Yarmouth and Ipswich in East Anglia lack the industrialized hinterland and are general cargo and fishing ports which now also serve the agricultural trades and the southern North Sea oil installations. Felixstowe is Britain's fastest-growing port, and is now the country's largest container complex.

The four Humber ports – Hull, Grimsby, Goole, and Immingham – have a location on the east coast similar to Merseyside's and Manchester's on the west coast. Hull and Grimsby compete for the position of the country's leading fishing port, but the former has diversified and now serves the general cargo trade as well as supporting a large petrochemical installation and roll-on/roll-off ferries. The developments at Immingham and Killingholme, in the deepwater channel on the south side of the estuary, specialize in petrochemicals whilst Goole was developed as the trans-shipment port at the estuary mouth of the inland canal system.

Shoreham and Dover, like Southampton are on the Channel coast and large enough to warrant inclusion in Fig. 8.2. These were all initially developed as passenger and ferry ports; the first two have continued to grow, serving continental traffic, whilst the last, which served the ocean liners of the first half of the century, has now lost its ferry traffic to Portsmouth. The demise of ocean passenger traffic with the development of air routes has forced Southampton to diversify and it now handles a large general cargo, petrochemicals, and is home port for many of the aggregate mining ships detailed in Chapter 11. Farther west on the Channel coast Plymouth developed as a naval dockyard around its magnificent natural harbour from earliest times and handles a cross-Channel ferry trade, whilst Par is almost exclusively concerned with exports for the china clay mines of Cornwall. Finally, the port of Bristol some 15kms up the tidal Avon from the Severn Estuary and the Bristol Channel became prosperous in the eighteenth and nineteenth centuries serving the whole of western England and offering direct access to the North Americas. Its merchants traded widely in all manner of commodities, including slaves, but the industrial revolution tended to by-pass the port, and the modern trade is centred on grain and petrochemicals through the opening of Avonmouth docks farther downstream.

Scotland

The Clyde is no natural waterway, although it was for centuries the world's busiest shipbuilding river. The major Clyde port of Glasgow was developed artificially with the growth of commerce at the end of the eighteenth century by narrowing the river to induce natural scouring, and by blasting away outlying rock barriers. The port was, until recently, Scotland's largest and primarily serves the industrial central lowlands, both exporting the products, and with the depletion of local reserves, importing much of the raw material requirements. Hunterston, also on the Clyde, is now the major ore terminal. The largest port in Scotland, indeed in the whole of the British Isles, is now Sullom Voe in the Shetland Islands, which has been developed exclusively to service the oil industry in the northern North Sea

(Chapter 9). The east coast of the country is more heavily populated than the Highlands to the west, and three ports are sufficiently large to warrant inclusion in Fig. 8.2. Grangemouth is conveniently located between the two cities of Glasgow and Edinburgh and also serves the heavily industrialized central lowlands, the oil industry and as Scotland's lumber port. Leith, the port of Edinburgh, carries more general cargo but also specializes in the receipt of grain. Finally, Aberdeen, farther north, is Scotland's third city and its premier fishing port. Catches from the northern grounds are rapidly transferred south by rail to supply the Scottish and British conurbations. The port has also seen much recent development in servicing the North Sea oil industry.

Wales

All of the major ports in Wales are located along the south coast on the deep water of the Bristol Channel, stretching from Newport, in the east, through Cardiff, Port Talbot to Swansea and the new petrochemical complexes at Milford Haven in the west. Both Cardiff and Swansea were developed as coal exporting ports from the middle of the last century and were served by canals, tramways and later railways from the mines in the Welsh valleys. The coal trade declined rapidly after the First World War, and the ports have made tremendous efforts to attract alternative trade. In particular bulk materials for the steel industry, which is now centred around Port Talbot, have been imported and exports of the manufactured goods are developing. Milford Haven is a large natural harbour with almost 20m of water at all states of the tide. It has been developed by the oil companies to provide both terminal and refinery facilities.

Northern Ireland

The concentration of economic activity around Belfast is striking, and this has resulted in the port carrying more than two-thirds of Northern Ireland's seaborne trade in a normal year. The trade is dominated by imports, largely from the mainland UK to deal with grain and feedstuffs for agriculture together with raw materials for the dominant linen, aircraft, and shipbuilding industries. The port has been carved out of the banks of the Lagan river estuary, and much of the traffic is handled by liners and ferries. However, change is afoot and containerization came early to these routes. Northern Ireland's first refinery at Dufferin Dock at the entrance to the port of Belfast was opened in the mid 1960s.

131

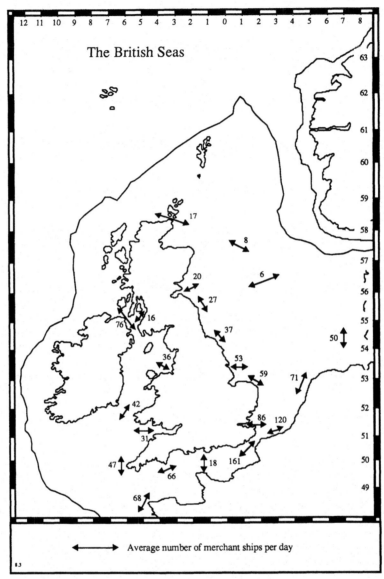

Figure 8.3 Traffic flow: merchant ships in the British Seas, showing average number of ships per day.

Source: Lee and Ramster (1981).

Eire

Dublin handles two-thirds of the Republic's imports by value and half of the exports, and is the focal point of the island's road, rail, and canal systems. The modern port has been created by constraining the River Liffey and by carving out docks and basins. Receipts are dominated by petroleum products and coal, livestock and Guiness ales are the major shipments from the port. The deep-water port at Bantry Bay has been developed in the last couple of decades, largely as a trans-shipment point for the largest tankers to discharge onto smaller ships for other European destinations. This trade has, however, declined with the changing structure of the world tanker fleet, and as a result of a serious accident when the *Betelgeuse* exploded in the harbour. The only other major port is Cork, which serves almost all of the country's industry with the exclusion of agricultural concerns. The two sections of the port, Lower Harbour which is a commodious natural bay and Upper Harbour which has been created within the city itself, therefore handle raw materials for the steel and rubber industries. Its position also made Cork into a usual port-of-call for Atlantic mail and passenger shipping, though this trade has ceased with the development of air transport.

Traffic flow

The passages and routes of both coastal and ocean-going merchant shipping around the British Seas will clearly reflect primarily the economic prosperity of the ports described in the previous sections, but will also be influenced by the demand to minimize transit costs and time, and by the navigational problems posed by the region. The major merchant shipping and ferry routes are shown in Figs 8.3 and 8.4, and will be discussed in this section.

Merchant ships

Fig. 8.3 gives an idea of the number and general location of merchant ships, other than ferries, to be found on any one day in the British Seas. It is based upon a study of 25,291 ship movements during two weeks in July 1976 (Lee and Ramster, 1981). The study utilized 26 traffic survey areas and 14 arrival or departure sectors in the Western Approaches. The 1976 findings were confirmed by samples from Lloyd's List between May 1976 and May 1977 and by further sampling in one area during 1978–9. Lee and Ramster (1981) note rather enigmatically that whilst at any one time the data show that a consid-

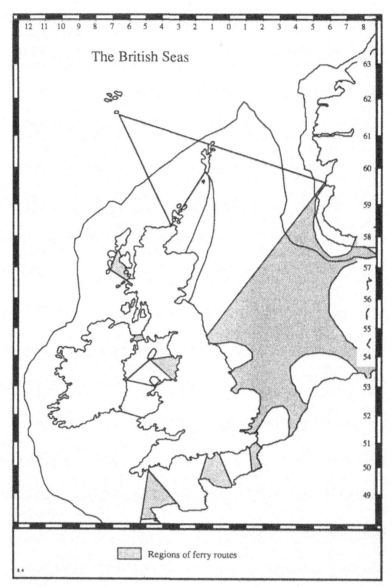

Figure 8.4 Traffic flow: regular seagoing ferry routes in the British Seas during 1979–80.

erable number of ships do ply the British Seas, it remains true that a few hours on passage towards, for example, the central North Sea, will bring the vessel to a region where, at least as far as the eye can see, the sea is empty for several days at a time, and this sense of isolation remains for some marine scientists one of the attractions of the subject.

Ferry routes

Fig. 8.4 was compiled from Anon. (1979) and illustrates the most regular sea-going ferry routes, including freight and train–ferry services, hovercraft and hydrofoils. Comparison with the merchant ship tracks in the previous section shows that, for the most part, ferry routes lie across the main lines of merchant ship movements, with the Strait of Dover providing an example of the extreme development of this trend. This general situation, together with the development of the oil and gas fields in the North Sea (Chapter 9) and the consequent arrival of drilling and production of all kinds, has led to attempts being made to lay down the routes to be followed by merchant ships in the region. These developments are discussed later.

Cable routes

In addition to the trade and shipping routes described in the preceding sections, the British Seas are criss-crossed by many hundreds of miles of submarine cables. These link the UK with offshore islands, with Europe, and with the American continent. The cables are generally multicore copper conductors and include signal amplifiers at regular intervals. A single cable can carry thousands of telephone calls, as well as Fax, Telex and data transmission, and, in some cases, television channels. There is presently an increasing development and use of fibre-optic cable, and with it a gradual change to digital rather than analogue transmissions.

The British cables are laid and maintained by the Post Office and by British Telecom. A fleet of purpose-built cable-laying ships is used to service the facilities and to repair the seventy or so cable breaks which occur each year. The cables (Fig. 8.5) can be categorized as:

1. Ten eastern Atlantic trans-ocean systems with a total cable mileage of 11,700 nautical miles.
2. Two hundred and eighty island and inter-island systems with a total mileage of 900 nautical miles.

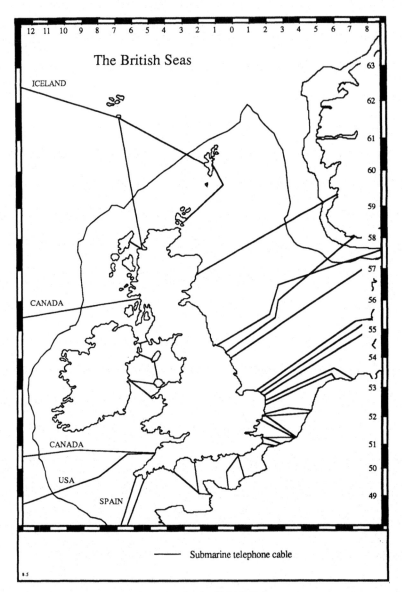

Figure 8.5 Submarine cable routes in the British Seas.

Source: Lee and Ramster (1981).

3. Twenty-eight European systems with a total mileage of 900 nautical miles.

In general the cable tracks follow the shortest direct route but each new cable is preceded by a detailed route survey and planning exercise. The objective is to balance a desire for minimal cost with the avoidance of natural obstacles and other potential problems such as tidally scoured outcrops, active sand wave fields, and fishing and anchoring grounds. In deep, quiet water unprotected cable is simply laid on the ocean floor, but the presence of potential hazards demands the use of expensive armoured cable or burial techniques (Davies, 1979).

Distribution of trade and shipping

This chapter has been concerned with the utilization of the British Seas as a transportation resource, for the movement of both cargoes aboard merchant shipping and information through submarine cables. This section attempts to utilize the spatially restrictive analysis which was introduced in Chapter 1 to draw broad conclusions about the density of such traffic in the British Seas.

Environmental restrictions

The whole of the British Seas is, in principle, open for utilization by merchant ships. However, the problems of navigating safely around complex coastlines have restricted the passage of (particularly the larger) bulk carriers to deeper water further from the shorelines (Chapter 2). Most ship tracks therefore tend to take the vessel offshore before turning to run along the coasts or towards the destination. A secondary environmental consideration is that when entering or leaving the ports, passages are generally timed to coincide with the tidal cycles discussed in Chapter 5. The reasons for this are twofold: first, the ships attempt to utilize flood and ebb currents to minimize journey times, whilst second, it is often only possible to enter coastal regions or the ports themselves at times close to high water. Both of these considerations mean that the timing, if not the actual routing of passages is affected by tidal conditions. A third environmental consideration is that inclement weather, and particularly adverse wave conditions (Chapter 4) frequently cause vessels to be delayed or re-routed, and ships' tracks tend to avoid the more exposed regions of the British Seas if all other factors can be equated. Some attempts at controlled ship routing to minimize the

The resources

effects of adverse wave conditions are reported by James (1957), but these do not appear to have been applied to the British Seas. In conclusion then the environmental potential for merchant shipping in the British Seas is limited by three factors: the necessary depth of water for safe navigation, the effect and timing of tidal movements, and the effects of adverse wave conditions.

Economic restrictions

This book cannot enter into a detailed description of the economic restrictions on trade between the ports in the British Seas, or between those ports and other, international destinations. The reader should refer to more specialized texts such as O'Loughlin (1967) for a general overview or Hoyle and Hilling (1984) for an analysis of port economics. We are instead interested here in the distribution of shipping within the British Seas. It is axiomatic that a ship will be found in transit along one of the routes shown in Fig. 8.3, provided that there is an economic demand for carriage between the port of loading and the destination. That demand depends upon the industrial hinterland serviced by the port, as described earlier, and upon the facilities for, and cost of, cargo handling. Once at sea the ship will take the shortest (and therefore cheapest) safe route, with the proviso that many vessels carry multiple cargoes necessitating less than direct routing. The economic restrictions are therefore largely the market forces of freight rates and demand.

Policy restrictions

There are, at the present time, very few restrictions on shipping and cable routes in the British Seas. Ship routes are decided by individual navigating officers and, with the exception of certain special cases, will be as direct and therefore as cheap as is permitted by the above considerations. The special cases concern the imposition of separation zones at particularly congested crossing points in the region for the purposes of navigation and safety. There are presently eleven such zones in operation and these are:

1. Strait of Dover
2. Central Channel
3. Isle de Ouessant, off north-west France
4. East of Scilly Isles
5. South of Scilly Isles
6. West of Scilly Isles
7. Western St George's Channel

138

8. Eastern St George's Channel
9. Anglesey
10. South of the Isle of Man
11. North Channel

In general these involve restrictions which require vessels bound in opposite directions to take particular sea lanes which are separated by clear lanes. In conclusion then, shipping routes in the British Seas are not influenced by policy constraints except for a small number of navigational restrictions.

References

Anon., 1971. *Oxford Regional Economic Atlas of Western Europe*. Oxford University Press, London, 162 pp.

Anon., 1979. *European Ferry Services, 1979–80*. IPC Business Press, London (Published by 'The Motor Ship').

Alexandersson, G and G. Norstrom, 1963. *World Shipping: An Economic Geography of Ports and Seaborne Trade*. John Wiley and Sons, New York.

British Ports Federation, 1988. *Port Statistics 1987*. Government Statistical Service, London.

Bird, J.H., 1971. *Seaports and Seaport Terminals*. Hutchinson, London, 240 pp.

Couper, A.D., 1972. *The Geography of Sea Transport*. Hutchinson, London, 208 pp.

Davies, A.P., 1979. Submarine cable systems: a review of their evolution and future. *P.O. Electr. Engnrs J.*, 72(2), 87–94.

Hoyle, B.S. and D.Hilling (Eds), 1984. *Seaport Systems and Spatial Change*. Wiley, Chichester, 481 pp.

James, R.W., 1957. *Application of wave forcasts to marine navigation*. U.S. Navy Hydrographic Office, Publication No. SP-1.

Kerchove, R. de, 1961. *International Maritime Dictionary*, 2nd Edn. New York.

Lee, A.J. and R.W.Ramster, 1981. *Atlas of the Seas Around the British Isles*. MAFF, Lowestoft.

Lloyd's Register, 1988. *Annual Report*. 33 pp.

O'Loughlin, C., 1967. *The Economics of Sea Transport*. Pergamon Press, Oxford, 218 pp.

Chapter nine

Hydrocarbons

In 1850 James Young, a Scottish chemist, devised a process by which paraffin wax, kerosene, and lubricating oil could be obtained from the natural oil which flowed into the workings of a Derbyshire coal mine, and soon afterwards the technique of boiling off the volatile constituents of the crude oil was sufficiently developed to obtain petrol. A quarter of a century later, in 1875, Nikolaus Otto constructed the first internal-combustion petrol motor and 10 years later Benz inaugurated the era of motor vehicles by fitting Otto's engine onto a wheeled chassis. Young's discovery continued to provide liquids from coal, and a number of small onshore gas and oil finds were made and exploited in the early decades of the twentieth century. In England these ranged from Eskdale in Yorkshire, through Gainsborough in Lincolnshire to Kimmeridge in Dorset. The pattern of widespread but small onshore hydrocarbon finds continued, and by the early 1960s total onshore oil production was running at approximately 1,500 barrels a day, which represented between 1 and 2 percent of the total UK oil requirements. However, the demand was also to grow remarkably, from 9.5 million tonnes in 1950, to 45 million tonnes in 1960 and to more than 100 million tonnes a decade later (Bending and Eden, 1984).

The pattern of a large number of small oil and gas finds was broken in 1959 when a major discovery of gas was made by the Shell–Esso consortium at Groningen in Holland which positively established the existence of major deposits of hydrocarbon fuels on the European continental shelf. The coastal nations had already anticipated this discovery, and had begun to place the exploration and exploitation of their coastal and offshore waters into a more rigorous legislative context. The process had begun through the Geneva Convention of 1958, which stipulated the basic guidelines to be used in determining the international boundaries on the seabed. The decision was that 'boundaries shall be determined by application of the

Plate 9 Offshore oil and gas installations must be able to operate in adverse environmental conditions (courtesy of Wimpol Ltd)

principle of equidistance from the nearest points of the baseline from which the breadth of the territorial sea of each state is measured' (Fig. 9.1). Although Denmark proclaimed sovereignty over its sector of the shelf in 1963, and the UK had anticipated the consequences of the convention by annexing Rockall in 1955, the convention was not subsequently ratified by most European countries until 1964, and thereafter offshore exploration began in earnest. The agreed seabed boundaries of the coastal nations are shown in Fig. 9.1, which largely reflects the results of the 1964 accord although the more distant areas had to wait until 1965 when agreements were reached between the UK and Norway, Germany, Denmark, and the Netherlands.

In the present chapter we review the technology of hydrocarbon extraction in the British Seas, and relate potential sites to the geological history of the region which was discussed in Chapter 3. The development of the gas and oil industries are then detailed separately, and the chapter concludes with an examination of the problems of pipeline routing across the mobile seabed in the southern North Sea (Chapter 7). An analysis of the industry is then used to examine the inter-play between, on the one hand, the geological and environmental controls on resource exploitation and, on the other, the effect of national and international policies on the development of the resource.

Hydrocarbon technology

The successful extraction of oil or gas from beneath the seabed requires the fortuitous combination of three different geological structures. First, there must be a suitable *source* rock, wherein temperature and pressure has converted organic remains into oil and gas. Second, since the hydrocarbons rarely remain in the source rock, there must also be a *reservoir* rock, wherein migrating hydrocarbon fluid collects in commercial concentrations in rock voids. Finally, because the low-density hydrocarbons would naturally continue to migrate towards the surface, there must also be a suitable *cap* rock, that is an impermeable layer over the reservoir which traps the deposit and prevents further migration. The technologies of the hydrocarbon industries are designed to exploit this threefold geological convenience. The technology is complex and costly for land-based fields, but the environmental and engineering problems are many times more so when deposits are found in the offshore regions. Principally the engineer must position a drilling platform on site, and once economic deposits have been proven he must bring them to the surface and transmit them to a shore-based refinery. The platforms used in the British Seas are detailed in the present section,

143

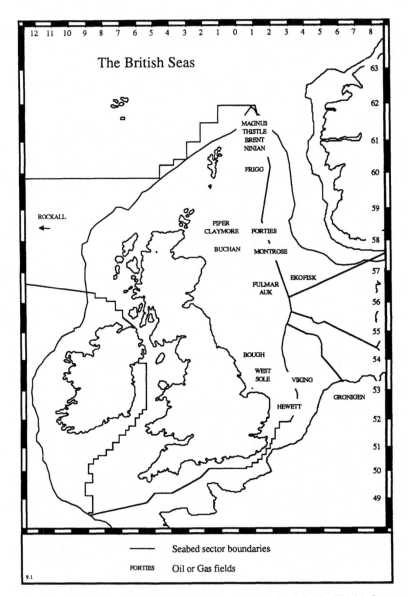

Figure 9.1 Seabed sector boundaries and oil and gas fields in the British Seas
Source: Lee and Ramster (1981).

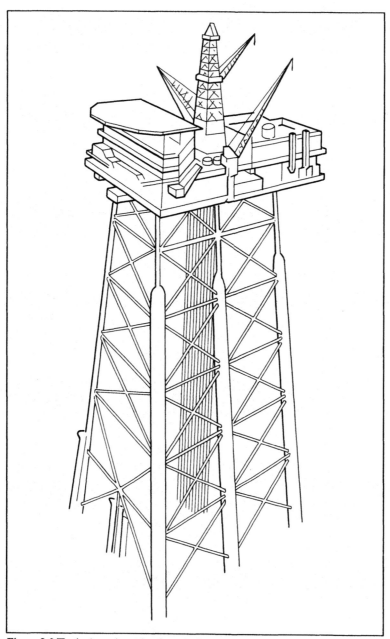

Figure 9.2 Typical steel production platform.

Figure 9.3 Typical concrete-gravity production platform.

and the pipelines are detailed later in this chapter.

There are two types of oil and gas platforms: first, there are the exploration platforms and second there are the production platforms. The exploration wells are drilled with the former type following geophysical surveys of an area. These rigs are mobile and relatively lightweight. More than 858 exploration and appraisal wells have been drilled in the North Sea (Klitz, 1980) of which only 144 are directly attributable to present-day production sites. The major engineering facility is, however, the production platform. The production wells are drilled from these, and are often non-vertical, deviating considerably with depth.

The design, construction and installation of a platform occurs over a period of 2 to 4 years. The North Sea platform is designed to withstand the type of storm which occurs once every hundred years (Chapter 4), that is a wave height in excess of 30m in the more exposed regions of the British Seas. There are two types of production platform currently in use in the North Sea: the steel jacket platform (Fig. 9.2), which is attached to the seabed through pilings, and the concrete-gravity platform (Fig. 9.3), whose enormous bulk allows it to remain stationary on the sea bed. An additional feature of the concrete-gravity platform is that it can provide offshore crude storage within the concrete structure itself. Steel platforms range from Beatrice in 45m of water, through the 6,800 tonne Auk in 74m to the colossal 84,000 tonne Magnus in 187m of water. Concrete-gravity platforms are necessarily heavier in order to remain stationary on the seabed and include the Central Ninian platform, weighing more than 450,000 tonnes and located in 140m of water.

A deck is fitted over the support structure to carry such equipment as the drilling rigs, production and process equipment, living quarters, pipeline pumps, recovery equipment, helicopter landing pads, and communications equipment. The directionally drilled oil-production wells sunk in the North Sea are generally from 3,000 to 4,500m deep, and on average about 12 weeks is required to reach such a depth. The casing which lines the well after retrieval of the drill pipe is usually also a steel tube, and this will weigh about 750 tonnes for a 3,500m well. It has been estimated that some 1,700 production wells will have been drilled in the North Sea by the end of the 1980s (Klitz, 1980).

Hydrocarbon geology

The geological history of the British Seas was dealt with in Chapter 3. This section reviews the elements of that history which are appropriate to the commercial exploitation of oil and gas reserves. The ultimate, technically recoverable hydrocarbon reserves in established accumulations in north-west Europe amount to 27.8×10^9 barrels of oil and 215×10^{12}s ft^3 of gas; and of this, some 3.2×10^9 barrels of oil and 50×10^{12}s ft^3 of gas have already been produced (Ziegler, 1982).

The bulk of these reserves are contained in the north-west European Basin (Fig. 9.1) which extends from the shelves of the Shetland Islands and western Norway to northern Germany and Poland. Major production has only been achieved in the northern part of the basin and here the reserves are contained in three distinct hydrocarbon provinces, two of which are located in the North Sea and one is on land in northern Germany. It is only the former two fields which concern us here.

The Permian gas province extends across the Netherlands and the southern North Sea and is essentially tied to the southern margins of the Southern Permian Basin (Chapter 3). Source rocks are provided by the Westphalian coal measures of the Variscan foredeep; but the gas has migrated and the main reservoir rocks are now the Rotliegend Sands and the Zechstein carbonates. The Permian gas province therefore relies for its hydrocarbon potential on the superposition of several genetically different basins to provide the necessary source, reservoir, and cap rocks for economic hydrocarbon production. The hydrocarbon reserves in the southern North Sea amount to some 145×10^{12}s ft^3 of gas (Ziegler, 1982).

The central and northern North Sea contains the most important hydrocarbon province of north-western Europe with reserves of some 24×10^9 barrels of oil and 55×10^{12}s ft^3 of gas (Ziegler, 1982). Even these figures require updating with the recent discoveries in the Norwegian sector of the Viking Graben. The main source rocks are of late Jurassic age and the reservoirs range from Devonian to early Tertiary. The present-day disposition of recoverable resources are mainly defined by the configuration of the Cenozoic rifting of the North Sea Basin (Chapter 3).

Additional reserves in the British Seas have also been identified in the Celtic Sea Trough, and the Manx-Furness, English Channel, and West Shetland Basins, although early estimates suggest that they will be dwarfed by the North Sea.

In many of the established hydrocarbon provinces exploration has already reached a mature stage. Remaining frontier areas are the deeper water parts of the northern North Sea, of the Atlantic Shelves of Ireland, Scotland, and western Norway, and the vast Barents Sea shelf.

Gas production

The rights to prospect for both oil and gas are allocated in exactly the same way. Licences for exploration and production are granted under the 1934 Petroleum (Production) Act and the 1964 Continental Shelf Act, which was itself amended by a number of subsequent sets of regulations. The licences are of two types: exploration licences which allow preliminary geological survey work and production licences which are necessary for any large-scale drilling activity. We shall see that a similar system operates for seabed mining (Chapter 11). The granting of a production licence does not, therefore, mean that oil or gas has positively been discovered, but only that deep exploration drilling may take place. The licencees who carried out the early gas exploration were, by and large, the established oil companies. This is illustrated by the companies which made the major gas finds: British Petroleum in West Sole, Shell/Esso in Leman Bank and British Gas/Amoco in Indefatigable.

The distribution of proven gas fields at the present time is shown in Fig. 9.1. Only one well was actually drilled in 1964 and this met with no success. In 1965 a further ten wells were drilled and the first significant gas find was made in the West Sole field. The peculiar economics of this product then began to influence both the exploration for, and exploitation of the North Sea gas reserves. In the UK almost all gas is supplied to the customer by what in the 1960s was called the Gas Council, and later became the British Gas Corporation and now simply British Gas. This company therefore operates a monopoly on the purchase of the product through its ownership of the distribution network.

Before 1965 almost all of this gas was produced from coal or oil as shown in Fig. 9.4(a). However, the potential size of the reserves led to an important policy decision to convert all the gas appliances in Great Britain to use natural gas. This was designed to stimulate demand for the product, which had almost twice the thermal equivalent of traditional gas. The policy argued that the additional demand was necessary in order to recoup the investment which had been made by the industry to convert all domestic and commercial appliances. Finally, the Gas Council decided to use its powerful position to establish the lowest possible gas price to the consumer. In February 1966 British Petroleum signed a contract to sell the natural gas from the West Sole field to the Gas Council for about 2 new pence per hundred cubic feet (£0.02). However, in 1967 the best offer made by the Gas Council for the purchase of natural gas was little more than a half a new penny per hundred cubic feet (£0.005). The policy was very successful and quickly led to a massive increase in the consump-

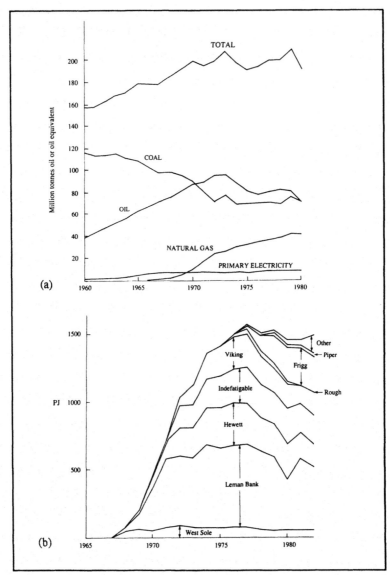

Figure 9.4 (a) UK mainland consumption of primary fuel for energy use, and (b) the build up of natural gas production.

Source: Data from Atkinson and Hall (1983).

Table 9.1 North Sea gas production and value

Year	Production (million tonnes oil equivalent)	Price (£ per tonne oil equivalent)	Value of gas (£million)
1967	0.39	10.9	4.3
1968	1.75	8.6	15.1
1969	4.39	6.2	27.2
1970	9.62	5.4	51.9
1971	15.62	5.3	82.8
1972	22.61	5.1	115.3
1973	24.62	5.7	140.3
1974	29.74	6.6	196.3
1975	30.98	7.6	235.4
1976	32.83	8.6	282.3
1977	34.44	12.0	413.3
1978	32.9	18.7	615.2
1979	33.52	22.8	764.3
1980	31.87	23.6	752.6

tion of gas in the UK, as shown by Fig. 9.4(b). The corresponding production figures and price structure for all of the North Sea gas fields are shown in Table 9.1.

By the late 1970s five gas fields were producing the majority of the gas from the North Sea as shown in Fig. 9.4(b), but it is apparent that production from Leman, Hewett, Indefatigable, and Viking was beginning to be reduced, and the Frigg field was being worked to generate new supplies. This was to be expected from the increased demand generated by the policy of British Gas to provide the product at the minimum possible cost to the consumer, and the future of the industry depends upon further fields being brought on line.

One final operating problem for gas extraction is that domestic and commercial heating is subject to extreme seasonal variations. These represent a major factor in the demand for natural gas, and the economic penalties of imposing these variations on the offshore production wells are substantial. It is therefore desirable to provide shoreside storage or to make other special arrangements to accommodate these fluctuations. Two salt cavities at Hornsea in Humberside, close to the Easington terminal for the Rough and West Sole fields, were opened in 1981 and have a capacity of 2,000 million cubic feet. This is useful on a daily cycle basis. Additionally the relatively new Morecambe Bay field, which was opened in 1974, is being developed specifically to provide seasonal supply, and consideration is being given to the use of the depleting southern sector fields for offshore storage.

Oil production

The early offshore oil discoveries were made in the Central North Sea Basin (Fig. 9.1) which extends up to the two small fields Tiffany and Toni. The area to the west of this is the Moray Firth Basin and to the north the East Shetland Basin. Virtually all the discovered offshore reserves are contained within these three basins. Exploration in these areas had increased rapidly from ratification of the Continental Shelf Act and the award of the first licences in 1964, but stabilized in 1969 and then dropped dramatically. However, this pattern changed in the early 1970s with two separate developments. First, the discovery of major oil fields established the existence of truly commercial deposits in the North Sea. The Forties field was found in November 1970 and is estimated at 240 million tonnes. The Brent field was discovered eight months later in July 1971, and this is estimated at 229 million tonnes. Second, the sudden increase in the world price of oil occurred at this time, resulting from the closure of the Suez Canal and the increased western dependence on Libyan oil. When, in 1970, a left-wing Government came to power in Libya, they immediately reduced oil output and increased the price, other producing countries quickly followed suit. The result was that, in 1973/74, oil prices rose from $2.80 to $11.60 a barrel. The North Sea hydrocarbon economics quickly changed from being a marginal operation to one of lucrative profits, and Table 9.2 shows the resulting accelerated development rate for the UK sector in comparison with demand.

In 1979/80 the UK became a net exporter of oil, and this obviously had a dramatic, though temporary effect on the country's balance of payments. Additionally the early legislation which divested ownership of fossil fuels in the Crown Estates, and led to the licensing procedures outlined earlier, permitted the government to devise and implement a taxation structure on the production companies which generated considerable revenues for the exchequer. This should be set against the capital investment required to recover the product from the harsh offshore environment, and the direct operating costs of the producers. As shown in Table 9.2, early production was expensive, but became rapidly cheaper as output rose in 1977, then remained steady throughout 1978/79. In the 1980s the average cost began to rise quite sharply as the smaller, less profitable fields such as Murchison came onstream and production began to fall in some of the older fields such as Piper, Brent and Auk. The average cost of production is likely to continue to rise as smaller fields are exploited and the output of larger fields begins to decline.

Table 9.2 Oil production from the UK sector of the North Sea, UK oil consumption, capital investment and revenue

Field	Production (million tonnes)						
	1975	1976	1977	1978	1979	1980	1981
Argyll	0.5	1.1	0.8	0.7	0.8	0.5	0.5
Auk		1.2	2.3	1.3	0.8	0.6	0.6
Beatrice							0.2
Beryl		0.4	3.0	2.6	4.7	5.4	4.7
Brent		0.1	1.3	3.8	8.8	6.8	11.1
Buchanan							0.9
Claymore			0.3	3.0	4.0	4.4	4.5
S.Cormorant					0.04	1.1	0.7
Dunlin				0.7	5.7	5.2	4.7
Forties	0.6	8.6	20.1	24.5	24.5	24.6	22.8
Heather				0.1	0.8	0.7	1.2
Montrose		0.1	0.8	1.2	1.3	1.2	1.1
Murchison						0.4	3.1
Ninian				0.04	7.7	11.4	14.3
Piper		0.1	8.0	12.2	13.2	10.4	9.8
Statfjord					0.04	0.5	1.2
Tartan							0.7
Thistle				2.6	3.9	5.3	5.5
Total production*	1.6	12.1	38.3	54.0	77.9	80.5	89.4
Total consumption	93.3	92.5	92.9	94.0	94.0	80.8	74.4
Expenditure (£million) 1975 constant prices	1371	1843	1699	1577	1294	1229	
Government revenue (£million)		81	238	526	2324	3840	6430
Operating Costs (£ 1980/tonne oil)		11.7	6.3	6.5	6.3	7.4	8.7

* Including onshore production.
Source: Atkinson and Hall, 1983.

Pipeline routing

There are two alternative methods for transporting oil and gas from the fields to the coast: either it can be loaded into tankers at sea via several types of offshore loading terminal, or it can be piped ashore. In the case of gas in large quantities, pipelines are the only practical method. A large number of combinations are employed in the transport of oil in the North Sea fields. In the initial stages of field development, tankers are loaded either directly from the platform or

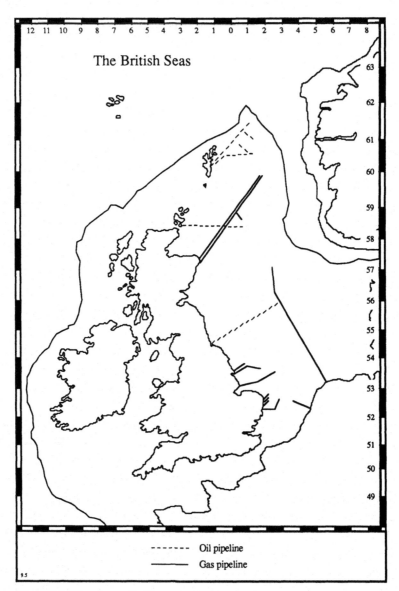

Figure 9.5 Oil and gas pipeline routes.
Source: Lee and Ramster (1981).

from adjacent loading facilities. Once a particular offshore pipeline is constructed to serve several fields that are in different stages of development, the offshore terminal is removed and the oil and gas is transferred ashore through the pipeline. In those cases where production does not warrant a pipeline, transfer is made via tankers with some offshore storage capacity being provided to match production to the rate of removal by tanker.

The amount of offshore storage in the North Sea varies widely between the different fields. For those fields (such as Argyll) where the buoy mooring systems are used for direct tanker loading, no offshore storage exists, whereas the massive concrete-gravity platforms (such as Brent B and Brent D) have internal storage of up to 1.1 million barrels in the platform itself (Klitz, 1980).

There are two systems of offshore pipeline, intra-field pipelines and inter-field pipelines. Intra-field pipelines transmit crude oil and/or gas to offshore loading terminals, to other production platforms, or from subsea wells to production platforms. Such pipelines range in size from 4.5 to 24-inch outside diameter. The inter-field pipelines, which are generally larger (16–36-inch outside diameter), transport the resource from one field to another or from the field to terminal facilities ashore.

The chart in Fig. 9.5 shows the pattern of pipelines which have been laid on the surface or buried beneath the seabed to bring the oil and gas ashore. In general, it is apparent that each major field has a series of short intra-field pipelines, which carry the product to a centralized location, from where a single inter-field pipeline connects to the shoreside installations. Thus the Brent field connects with terminals on the Shetland Islands, the Forties fields with Aberdeen, the Ekofisk fields with North Shields, and the gas fields in the southern Basin with either Easington on Humberside or Yarmouth in Norfolk. The fact that the long pipeline from the eastern Ekofisk fields in the Norwegian sector runs to Germany illustrates the enormous technical problems which would be involved in laying a pipeline and pumping oil or gas across the deep Norwegian Channel (Chapter 2). However, two routes are now being developed from Frigg and Brent directly to Norway, although the technical problems of working in such deep water remain immense. More than 3,500km of oil and gas pipelines have been commissioned in the UK sector of the North Sea up to the present time.

The laying of an offshore pipeline is a demanding task. The pipeline is first coated ashore with a mixture of cement, iron ore, heavy aggregates and steel reinforcement to protect and ballast the pipework. The pipelines are loaded onto a transport ship or barge in 12 or 24m sections and are assembled and welded together on a surface

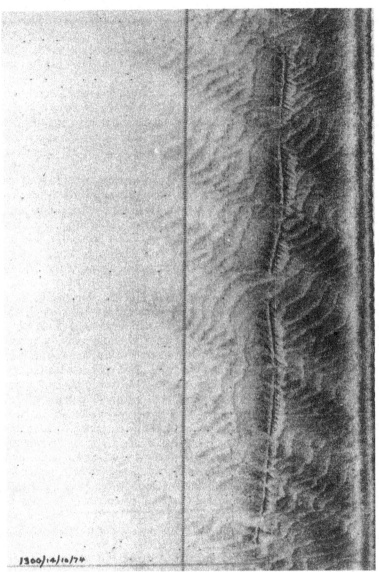

1300/14/10/74

Figure 9.6 Side-scan record showing pipeline scour.

Source: Photograph provided by N. Kenyon.

barge into a long 'string'. The string is then lowered to the seabed to remain resting on the surface, or to be trenched into place. The problems of such deployments, particularly in the southern North Sea, are illustrated in Fig. 9.6 which shows a pipeline crossing a sand wave field off Holland. The pipe is suspended between the crests and is dangerously exposed to flexure and breaking with each run of the tide. Additionally, scour can erode base supports, leading again to vulnerable weakenings of the integrity of the structure. The mobile sand wave fields of the southern North Sea (Chapter 7) are particularly difficult for these reasons, and the engineers attempt to use side-scan sonar mapping to lay the pipes along sand wave troughs, or become involved in the very considerable extra expense of burying pipes beneath the seabed.

The distribution of hydrocarbons

This chapter has been concerned with the utilization of the British Seas as a hydrocarbon resource. This section attempts to utilize the spatially restrictive analysis which was introduced in Chapter 1 to draw broad conclusions about the distribution of the industry in the British Seas.

Environmental potential

The geological evolution of the British Seas has resulted in a rich, though spatially restricted hydrocarbon resource. The most important areas are the Permian basins of the central and northern North Sea. The gas reserves are located along the southern margins of the Southern Permian Basin, where the source rocks are the Westphalian coal measures. The gas has, however, migrated and the reservoir rocks are the Rotliegend Sands and the Zechstein carbonates. The oil reserves are located further north, where the source rocks are of late Jurassic age, and Cenozoic faulting has trapped migrating oil in reservoirs of Devonian to early Tertiary age. There is the potential for further hydrocarbon recovery in the Celtic Sea, and in the Manx-Furness, English Channel, and West Shetland Basins at some future date. In conclusion then, the environmental potential for hydrocarbon exploitation in the British Seas depends upon the fortuitous combination of suitable source, reservoir, and cap rocks; and this combination has been well-proven in the North Sea basins.

The resources

Economic restrictions

The development of the offshore oil and gas industries involves colossal investments. Table 9.2 shows that almost £10 billion were
required during the development years from 1975 to 1980. Clearly
the international oil and gas companies cannot contemplate such
sums unless there is a measure of certainty that appropriate returns
will be generated on the capital. There are, of course, two factors to
the equation which provide economic restrictions on the resource.
First, as we have seen, the market price of the oil and gas must be sufficiently high to generate the required returns. This price fluctuates,
and exploitation from the British Seas depends upon long periods of
relatively higher prices. Second, the offshore extraction of hydrocarbons becomes more expensive if the fields are smaller, or if the
environment is harsher or further from land. Thus economic restrictions have concentrated extraction in the large fields, and shallower
waters of the North Sea. It is not, however, inconceivable that continuing world price rises, and the depletion of the North Sea reserves,
will lead to development of the smaller fields, and of regions in
deeper water, such as the Atlantic shelves of western Ireland, Scotland and Norway.

Policy restrictions

Policy restrictions have not, as yet, seriously discouraged the oil and
gas companies from removing hydrocarbons from any region within
the British Seas. The system of exploration and production licensing,
and of taxation on revenue has been designed to encourage ongoing
exploitation of the resource. It has, on the whole been successful in
this objective, and has also generated large sums for the revenue as
was shown by Table 9.2.

In conclusion, the distribution of the hydrocarbon industry in the
British Seas is dominantly controlled by the geological structures,
and to a lesser extent by the cost of extraction from more hostile,
deeper waters. The sale of oil and gas is a business, and the colossal
investment has only been made because rising international oil and
gas prices increased the probability of a return on capital.

References

Atkinson, F. and S.Hall, 1983. *Oil and The British Economy*. Croom
   Helm, London.
Bending, R. and R.Eden, 1984. *UK Energy: Structure Prospects and
   Policies*. Cambridge University Press, Cambridge.

Clark, R.B., 1988. *The Waters Around The British Isles*. Clarendon Press, Oxford.

Klitz, J.K, 1980. *North Sea Oil*. Pergamon Press, Oxford, 260 pp.

Lee, A.J. and J.W.Ramster, 1981. *Atlas of the Seas Around the British Isles*. MAFF, Lowestoft.

Ziegler, P.A., 1982. Evolution of sedimentary basins in North-West Europe. In: Illing, L.V. and Hobson, G.D. (Eds) *The Petroleum Geology of the Continental Shelf of North-West Europe*. Academic Press, London, pp. 3–39.

Chapter ten

Fishing

The offshore UK fishing industry (Fig. 10.1) began to develop when good catches were made in the North Sea grounds in the nineteenth century, and rich areas were named with timely analogies to gold rushes as in Klondyke, California Ground, and the Silver Pit. The long-distance whaling industry also developed at this time with voyages lasting for 2 years or more to the northern seas and to the southern oceans.

Steam and later diesel replaced sailing vessels from the turn of the century but until the 1950s position fixing (Chapter 2) was by cross bearings within sight of land, and by dead reckoning and testing the bottom with lead and line when offshore. Today, fishing at specific locations occurs along radio bearings, along depth contours and along Decca navigator co-ordinates. In the case of fishing for pelagic species (those swimming away from the bed) and to a lesser extent for demersal species (those swimming close to the bed), it is also commonplace to use sonar techniques to locate and identify shoals. Together with a knowledge of wrecks, and of how the seasons, the tides and the weather affect the fish and the set of the gear, these are the secrets of the trade which is currently landing more than a million tonnes of catch worth some £760 million pounds in the UK each year.

The present chapter begins with a general introduction to marine life, and in particular to the commercial pelagic and demersal fish species. These sections draw extensively on the oceanographic results which were presented in Chapter 6 in order to explain the distribution of fish spawning and feeding grounds in the British Seas. The chapter then considers the technology and distribution of the modern fishing fleets, and concludes by again analysing the relative environmental, economic and legislative influences on the distribution of fishing across the region. The following is based upon King (1975), Lee and Ramster (1981) and upon the Sea Fish Industry Authority (SFIA) Annual Report (1987), together with other more

The resources

Plate 10 Sampling for the phytoplankton, which form the base of the marine food chain (courtesy of NERC)

162

specific references on particular aspects of the industry.

Marine life

Plants are the essential basis of life both on the earth and in the
oceans. They alone can synthesize living matter from the chemical
nutrients in seawater.

The process is achieved with the aid of light derived from the sun,
by photosynthesis. Animals can then live on the plants, and flourish
in the ocean in an infinite variety of forms, each adapted to deal with
the special conditions of its environment. Plants can therefore only
exist within the layer into which the sun's rays can penetrate, and this
is known as the *photic zone*. Although the photic layer may extend to
more than 100m in the clear open ocean, it is restricted to less than
40m in temperate seas on continental shelves (such as the British
Seas) and often to much less near the coast because of the presence
of suspended material in the water column.

The term phytoplankton (from the Greek *phyton*, a plant and
planktos, wandering) is applied to the minute plants drifting at the
mercy of the surface currents on which all of the other marine crea-
tures depend for their living. Temperature, salinity, light, and the
presence of nutrients are all important for the production of phytoplankton. Temperature affects their rate of growth and of
reproduction, which in general declines if the temperature falls. High
temperatures and low salinities also decrease the density of seawater,
which makes it more difficult for the plankton to keep afloat in the
upper, photic zones.

The most important control on the production of phytoplankton
is, however, the presence of the nutrients, such as the phosphates, ni-
trogen, silicon, copper, and iron. The nutrients are present in weak
concentrations in all seawater, but to be of use to the phytoplankton
these nutrients must be present in the uppermost layers, where there
is also light. Once the surface supply has been used by the plants, the
sea loses its fertility unless the nutrients are replaced from below,
where they are continually being generated by the decomposition of
dead organisms by bacteria. Therefore any process which can stir up
the water is important because it brings nutrient-rich water to the
surface. The waters of the North Sea illustrate the pattern as is shown
in Fig. 10.2. In winter the sea is well stirred by waves and currents and
nutrients are distributed throughout the water column. The weak
light, dirty water, and low temperatures, however, inhibit plankton
activity. In spring, reproduction starts as the waters warm (Chapter
6) and sunlight levels increase, and phytoplankton levels increase

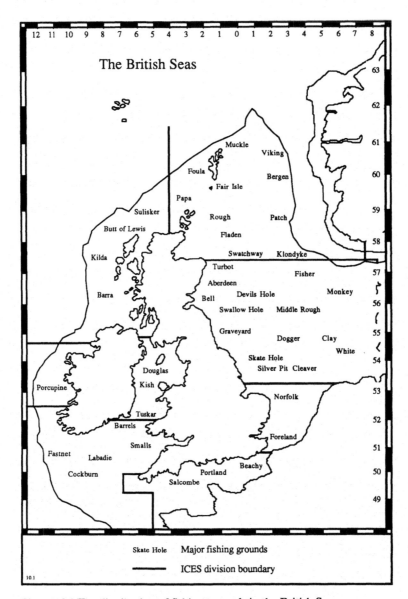

Figure 10.1 The distribution of fishing grounds in the British Seas.

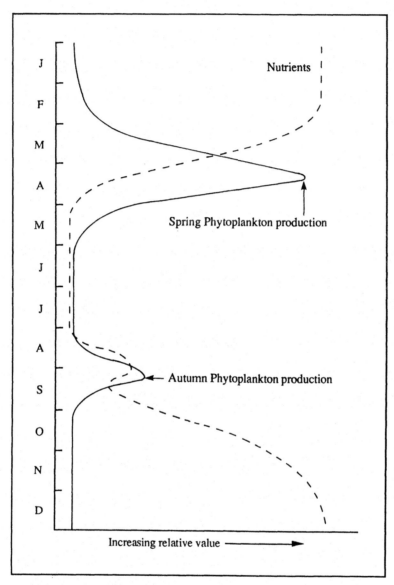

Figure 10.2 The yearly cycle of phytoplankton production in the North Sea and some factors to which it is related.

Source: After Lee (1958).

rapidly in a process which is known as the spring bloom. The next stage in the food chain then comes into effect as minute animals called zooplankton graze the crop of phytoplankton and the stratified waters can no longer replenish the nutrients. The phytoplankton concentrations therefore decrease until the winter storms again stir the waters in the less significant autumn bloom, before failing light and temperature levels again reduce activity to a low level.

In addition to these seasonal changes, the presence of oceanographic fronts (Chapter 6) marks sites where the interface between stratified and mixed waters also brings nutrient-rich waters to the surface, and results in concentrations of phytoplankton. The following sections discuss these effects in relation to the commercial fish species in the British Seas.

Species of fish

The phytoplankton form the primary producers in a complex food chain. The zooplankton convert the plants into the higher animals which ultimately form the food for various species of fish. Commercially exploited fish are divided into three categories: those which are used for human consumption, those which are used in industrial processes and for the production of fishmeal, and shellfish. The following details each category of fish, and then describes the geographical distribution within the British Seas of the main commercial species within each category.

(i) Fish which are used for direct human consumption:
   (a) Demersal species, which are fish living mainly on or close to the seabed and include cod, haddock, whiting, plaice, anglers, saithe, soles;
   (b) Pelagic species, which are fish living mainly away from the seabed and include herring, sprats and mackerel.
   These fish are illustrated in Fig. 10.3 and the main species are described in detail below.

(ii) Industrial fish, which are used for conversion by special factories ashore into fishmeal and fishoil, and include sand eels and pout. These are also augmented by older stocks of the category (i) fish above.

(iii) Shellfish, which are also used for direct human consumption and include crabs, lobsters, shrimps, oysters, mussels, and scallops.

The major commercially exploited species are the category (i) fish,

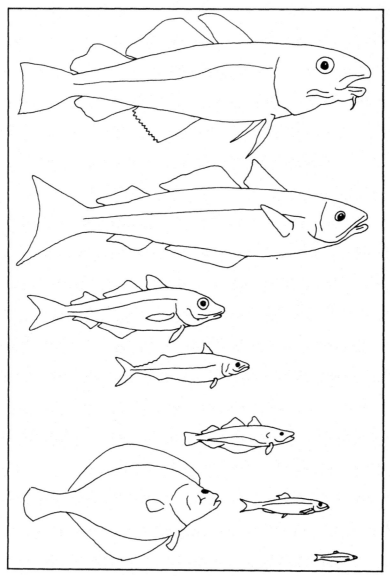

Figure 10.3 Commercial species (from top): cod, saithe, haddock, mackerel, whiting, herring, sprat, and (bottom left) plaice to same scale.

Table 10.1 Commercial demersal fish landings, 1986

Species	Kilotonnes	£ per tonne
Cod	76	905
Haddock	131	603
Whiting	40	453
Plaice	21	752
Anglers	7	1,683
Saithe	17	360
Dogfish	11	566
Dover sole	3	4,544
Sand eels	36	28
Others	33	1,007

Table 10.2 Commercial pelagic fish landings, 1986

Species	Kilotonnes	£ per tonne
Herring	106	112
Sprats	6	94
Mackerel	132	112
Other	6	63

those caught for human consumption. The demersal fish landings in this category by British vessels for 1986 are given in Table 10.1. The corresponding figures for the pelagic species are given in Table 10.2.

Some 87 kilotonnes of shellfish were landed during the same year and had an average value of £885 per tonne. The fishing industry is involved in an active marketing campaign to persuade buyers to take not only the existing, more commonly consumed species, but also more of the less common fish.

The following sections detail the distribution of the spawning and fishing areas for the most important species in the above tables. The details are taken from Lee and Ramster (1981) in which further information on the other species can also be found.

Demersal species

The demersal species (which live close to the seabed) were listed in Table 10.1. Three species – cod, haddock and whiting – account for more than two-thirds of the catch in this group and are detailed in the following sections.

Cod (Gadus morhua)

Cod is one of the most important commercial fish species in the region, and landings were, in 1986, worth more than £70 million
because of the relatively high price per tonne which the fish attracts.
Its distribution extends across most of the shelf, and there are also
important stocks at Faroe, Iceland, Greenland, and off the Atlantic
seaboard of North America. Long-distance migration appears to be
rare, and the annual migrations which do occur appear to be on a
relatively small scale. The adult fish tend to concentrate on the
spawning grounds in winter (Fig. 10.4) and to spread out over a wider
area during the summer.

Spawning is widespread and occurs mainly as the seas begin to
warm (Chapter 6) in the period from February to April. The eggs are
pelagic, as are the young fish for up to six months when the young
feed on crustaceans, worms, and small fish. Subsequently the young
fish adopt a more demersal habit. One-year-old cod are widespread,
but the main areas of concentration are in the Heliogoland Bight and
off the Danish coast. The fish mature at 3–5 years of age and are fully
grown at 6 years. Adult fish are carnivorous and feed on other fish.
There is a minimum landing size of 30cm enforced throughout the
region, except around the Faroes, where the limit is 34cm.

Cod are taken over the whole of the shelf, but mainly in the North
Sea, where the average annual catch rose from 122,000 tonnes in
1961–5 to 216,000 tonnes in 1971–5.

Haddock (Melanogrammus aeglefinus)

Haddock represents the largest of the demersal fish catches by weight
(Table 10.1), and despite attracting a lower price per tonne at market
than cod, the total value, in 1986, was more than £80 million. Stocks
are found from the Bay of Biscay to the Barents Sea on the eastern
side of the Atlantic. It is widely distributed in the British Seas, but is
not normally found in the shallower areas (Chapter 2) such as the
southern North Sea, the English Channel and the Irish Sea.

Spawning occurs in the northern North Sea and west of Orkneys
(Fig. 10.5) from March to mid-May and the eggs are planktonic. The
fish mature at the age of 2 or 3 years, and the adult fish feed by browsing on animals such as sea urchins, brittle stars, worms, snails, and
occasionally on other fish. They can live for 15 years and reach a size
of 90cm but 1–5 year olds form the bulk of the commercial catch.

A minimum landing size of 27cm is, however, enforced
throughout the region. Landings from the haddock fishery in the

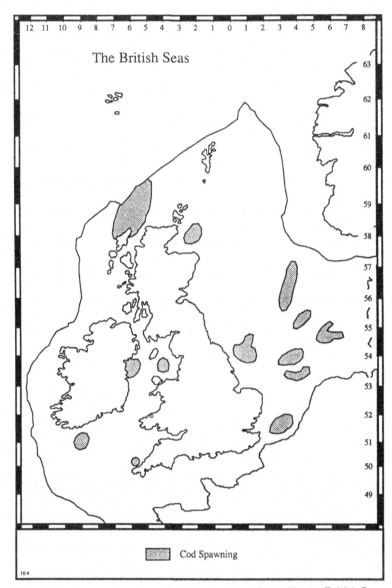

Figure 10.4 The distribution of spawning grounds for cod in the British Seas.

Source: Lee and Ramster (1981).

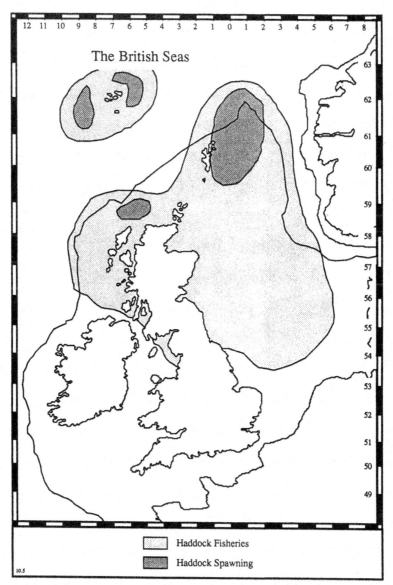

The British Seas

□ Haddock Fisheries

▨ Haddock Spawning

Figure 10.5 The distribution of spawning and fishing grounds for haddock in the British Seas.

Source: Lee and Ramster (1981).

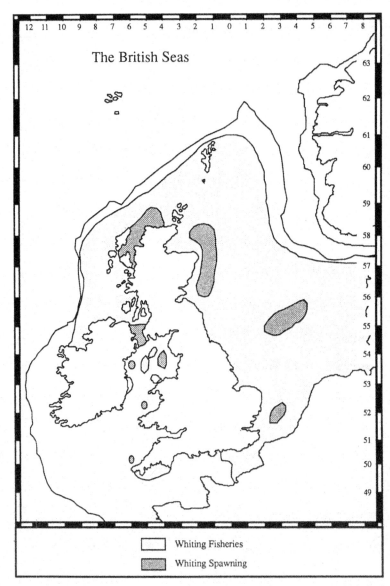

Figure 10.6 The distribution of spawning and fishing grounds for whiting in the British Seas.

Source: Lee and Ramster (1981).

North Sea fluctuate markedly because of extreme variations in the number of young fish. Record landings in 1972 amounted to some 670,000 tonnes but by 1978 this had declined to less than 100,000 tonnes.

Whiting (Merlangius Merlangus)

The whiting is the third most important species of commercial demersal fish in the British Seas. It is distributed widely in European waters from the Mediterranean and the Black Sea to the north coast of Norway and to Iceland. Consequently it is found over almost the whole of the continental shelf in the British Seas and the areas where it is caught by fishermen are similarly widespread, as shown by Fig. 10.6. It has been a particularly major component of the French fisheries for many years.

The food of whiting is mainly fish and crustaceans such as shrimps, with fish predominating in the diet of the larger individuals.

The spawning season is long, beginning in January in the south and extending northwards as the waters warm (Chapter 6). Spawning off east and western Scotland extends until August and September. Lee and Ramster (1981) note that information on the precise location of spawning grounds is limited and there may be others in addition to those shown in Fig. 10.6.

The maximum adult fish length is about 60cm, and the maximum age of the population is about 10 years. The fish first become mature adults at the age of 2 years and a minimum landing size of 27cm is enforced throughout the British Seas.

Much of the whiting stock caught in the North Sea fishery in recent years has been taken incidentally by fishermen hunting the industrial fishes, such as the Norwegian pout. Additionally large quantities of whiting have been discarded at sea in favour of higher-priced species.

Pelagic species

The pelagic species (which live within the water column rather than close to the seabed) were listed in Table 10.2. Three species, herring, sprat and mackerel account for more than 90 per cent of the catch in this group and are detailed in the following sections.

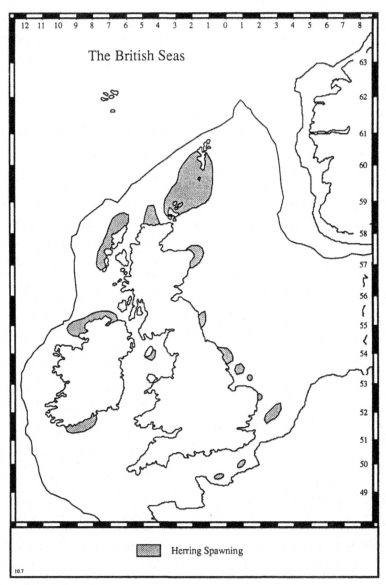

Figure 10.7 The distribution of spawning grounds for herring in the British Seas.

Source: Lee and Ramster (1981).

174

Herring (Clupea harengus)

Herring is widely distributed in the North Atlantic, and in European waters. The autumn and winter spawning grounds in the British Seas are shown in Fig. 10.7. Spring spawning herrings are also found in this region but their numbers are relatively very small. Herring eggs are laid on seabed gravels and thus closely follow the distribution of suitable substrates as detailed in Chapter 7. The directions of the prevailing currents were detailed in Chapters 5 and 6, and it is these which largely determine the position of the larvae after hatching has occurred. Thus larvae from the gravels off the Orkney and Shetland Isles are carried across the North Sea towards the Skagerrak and the coast of Denmark. Some of these larvae which hatch off the northeast coast of England reach maturity along the English east coast, whilst others which hatch further east are carried towards the German Bight and the west coast of Denmark. Larvae from the gravel banks in the southern North Sea and eastern English Channel are carried northwards along the Dutch coast towards the German Bight. Those hatching along the southern Irish coast tend to remain in that area, whilst those from the northern Irish Sea are carried towards the English coast between the Solway Firth and Liverpool Bay.

During the early months of life the young herring grow rapidly in the coastal waters to which they were carried as larvae. By late summer or early winter they will have reached a length of 8–14cm, and they begin to be less influenced by local currents, and to move away from the coasts into deeper waters. The feeding areas of the one year and older fish are the regions where zooplankton is abundant during the spring and summer months, and can thus be linked with the mixing described in Chapter 6. These feeding areas cover much of the British Seas region northwards from the southern Irish Sea, and the German Bight to the west and the east of the British Isles respectively.

However, this super-abundance has led to over-exploitation of the herring fishery. In the 1960s the North Sea catch alone averaged nearly 900,000 tonnes each year, and it was shared by all of the countries around the North Sea as well as by Russian, Polish, Icelandic, and Faroese ships. Over-exploitation led, however, to a rapid decline and from March 1977 North Sea herring fishing was banned. Similarly, bans are now in force to the west of Scotland and in the Celtic Sea. The Irish Sea herring fishery is strictly controlled at a level of about 10,000 tonnes each year.

Sprat (Sprattus sprattus)

The sprat is widely distributed along the Atlantic seaboard of Europe

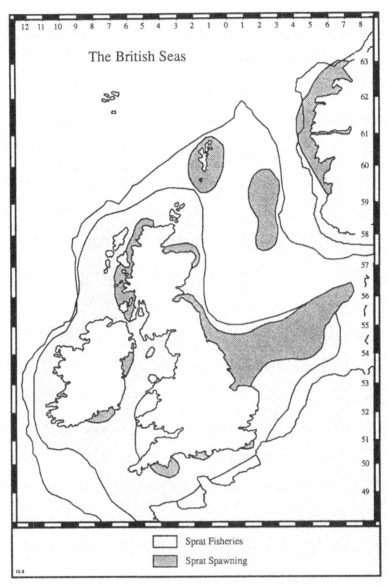

Figure 10.8 The distribution of spawning and fishing grounds for sprat within the British Seas.

Source: Lee and Ramster (1981).

176

extending from Portugal northwards to the Norwegian coast within the Arctic Circle. It is a shoaling fish which lives in surface waters feeding principally on planktonic copepods. In October and November when food becomes scarce it ceases feeding and lives on fat reserves. The fish shoal to overwinter in coastal waters where they become readily available for capture by trawls and purse seines (see techniques below). Most is industrial fishing for fishmeal, though small amounts are sold fresh and for canning and for pet foods.

Although sometimes coastal, the majority of sprat spawning occurs offshore and begins in the south becoming progressively later towards the north as the waters warm (Chapter 6). Peak spawning months are typically February–April south of Ireland and in the western English Channel, April–June in the southern and central North Sea and May–July in the northern North Sea. The eggs are planktonic and drift with the prevailing currents for 3 or 4 days before hatching. The larvae then live off yolk sacs for about a week, before they begin to feed on very small plankton. Sprats of 4–8cm length often drift into estuaries and embayments where shoals are called whitebait.

The adult sprat which usually spawns serially (i.e. eggs are released in discrete batches) from about 2 years old, eventually reaches a maximum length of about 17cm and has a lifespan of 5–6 years.

The fishing grounds (Fig. 10.8) show a remarkably close geographical correlation with the oceanic fronts described in Chapter 6, which is probably due to the presence of upwellings of nutrient-rich water, leading to the plankton on which these fish feed. The annual catches grew from about 30,000 tonnes per year in the 1950s to about 70,000 tonnes in the late 1960s and 1970s, and reached a peak of some 600,000 tonnes in 1975 due to the presence of Danish, Norwegian, and Russian boats. Most of the catch was taken from the central North Sea, tending to concentrate east of 3°E from July to November and west of 3°E from December to March. Smaller purse seine fisheries developed off Scotland and along the Channel, but the annual catches have been decreasing since 1975.

Mackerel (Scomber scombrus)

Mackerel in the eastern Atlantic occur from north-west Africa to central Norway, and the spawning ground in the British Seas is divided into two distinct stocks, each with its own overwintering area. The North Sea stock overwinters along the edge of the Rinne and the western stock overwinters in the Celtic Sea and along the continental shelf edge. As the waters begin to warm in spring and early summer

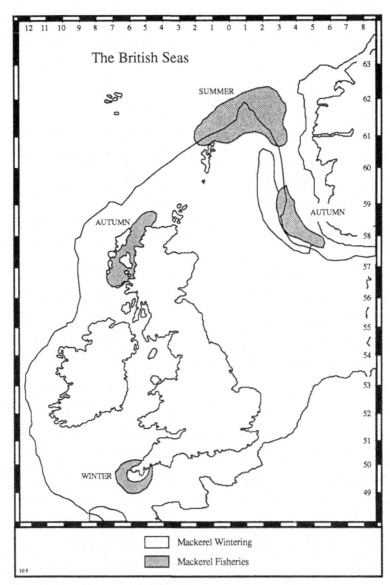

Figure 10.9 The distribution of spawning and fishing grounds for mackerel within the British Seas.

Source: Lee and Ramster (1981).

(Chapter 6), the mackerel move away from their overwintering grounds to spawn and feed as shown in Fig. 10.9.

The western stock spawns first. This is distributed along the continental shelf edge in the Bay of Biscay and Celtic Sea area in February–June, and then begins its feeding migration along the west of Scotland to the Shetlands and as far as the Norwegian Sea. Some of this stock also migrates eastwards along the English Channel and into the North Sea.

The North Sea stock does not appear to be so far ranging. It spawns in the central and northern North Sea in the early summer and then mixes with the western stock in the feeding grounds in the central North Sea and around Shetland. During winter the mackerel return to their overwintering grounds and appear to survive largely on fat accumulations. Lee and Ramster (1981) describe these seasonal changes in the geographical location of the mackerel but note that they are subject to some variability as the stocks respond to differing environmental pressures.

Mackerel eggs remain in the planktonic form for 5–10 days depending on the water temperature and then hatch as larvae and live off their yolk sacs before beginning to feed. By the beginning of their first winter the young mackerel are 12–18cm long and they remain more or less close to their spawning grounds until they are 3 or 4 years old. Historically the mackerel fishery has always been of secondary importance to the herring fisheries and it did not achieve significant levels until the herring stock began to be drastically reduced in the 1960s. The first fishery began to develop on the North Sea stock and landings increased from about 200,000 tonnes each year in the 1960s to over a million tonnes in 1967, taken largely by the Norwegian purse seine fleets. The stock could not withstand this intensity of effort and the fishery collapsed and there is still no sign of recovery in the stock.

Fishing on the western stock developed more recently, from about 100,000 tonnes in 1970 to 500,000 tonnes in 1976 and has developed off the Cornish peninsula in winter and off the Scottish coast in March. Thus the distribution of present-day Mackerel fisheries correlates closely with sea temperature but has been seriously affected by over-exploitation in the North Sea.

Fishing techniques

Although some commercial catches in the world fish market are taken with line fishing methods (Fig. 10.10), fishermen in the British

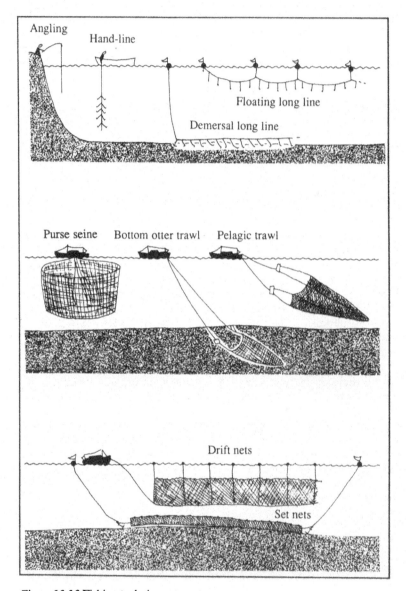

Figure 10.10 Fishing techniques.

Seas use one of four types of trawling or netting techniques. Demersal fish are taken mainly by otter trawls, beam trawls, and seine nets whilst pelagic fish are taken by purse seines and midwater trawls. Shellfish are taken by trawls, pots, and creels.

Otter trawl

Modern deep-sea trawlers use trawl nets in which the otterboard method of opening the trawl is used. The otterboards are large wooden or metal plates, frequently also known as doors, either rectangular or oval, attached to each side of the trawl opening so that they remain vertical and angled outwards. The pressure of the water on their surfaces as they are towed along forces them outwards and keeps the opening spread taut. A variation is the semi-pelagic or bottom trawl developed by Icelandic fishermen to catch fish, particularly cod, which normally lie just above the bottom. In this type of trawl the otterboards are towed along the seabed with the trawl bag floating a short distance above them.

Beam trawl

Before the development of the otter trawl, the beam trawl was the standard fishing gear for all deep-sea trawlers. In this type of trawl the mouth of the net is kept open by a wooden or metal beam 10–15m long which is attached to the top of the opening and mounted on skids at each end. The skids are designed to travel easily along the seabed. They are frequently used in pairs being towed one each side of the vessel and are nowadays used mainly by smaller vessels. Both otter and beam trawls gather their catch in the cod end which is opened on deck after recovery of the gear. Trawls can be operated in midwater for pelagic fishing by varying the speed of the vessel and shortening the warps.

Drift nets

Coastal boats use drift nets which are a long, shallow series of nets buoyed at the top edge with cork or with plastic or glass balls and weighted at the bottom with pieces of lead so that they hang vertically downwards in the sea and catch fish by the gills as they swim into the mesh.

Seine nets

This is a long, shallow net used in fishing for surface fish. The fish are caught within the net and not by their gills in the mesh. The purse seine net does not have a cod end but rather has a rope along the bottom edge so that the bottom can be drawn together after shooting to form a purse so that the fish enclosed cannot escape.

Fishing grounds

The commercial fishing fleet in the British Isles is divided into deep-water vessels (over 25m in length) and inshore vessels (12–25m in length). The deep-sea fleet numbered 205 vessels in December 1986, whilst the inshore fleet numbered 1,630 vessels. The ports of Hull, Lowestoft, Brixham, Plymouth, Milford Haven, Peterhead, Fraserburgh, and Shetland all had ten or more registered deep-water vessels and more than 16,000 fishermen were regularly employed at this time. The fishing grounds within the region are divided up, for statistical purposes, by the International Committee for the Exploitation of the Sea into six geographical areas, although the North Sea is often subdivided into northern, central, and southern sectors (Fig. 10.1). Average annual international catches during the period 1973–7 are shown in Table 10.3.

Table 10.3 Average international catches, 1973–7

Area	Kilotonnes (excluding shellfish)	Kilotonnes (shellfish)
North Sea	3151	218
North-western Approaches	416	13
Irish Sea	89	21
Celtic Sea	360	14
English Channel	207	84
West of Ireland	40	3

Fig. 10.1 shows some of the better known fishing grounds in the region. These names reflect the historical use of coastal landmarks for navigation, and the recognition that good fishing is to be found over particular banks or rough or fine grounds. The existence of a fishing port reflects three economic factors which are the historical or contemporary proximity to a locally dependent community, a fishing ground and a distribution or marketing outlet. The relative importance of the fishing ports in the British Isles will vary from year to year, but a typical table from the mid-1970s recognizes six groups, divided according to the annual value of the fish catch:

Catch (£millions)

0.1–0.5 Mevagissey, Looe, Plymouth, Salcombe, Dart, Portsmouth, Newhaven, Hastings, Leigh, West Mersea, Colchester, Brightlingsea, Wells, King's Lynn, Boston, Filey, Hartlepool, Sunderland, Blyth, Amble, Clogerhead, Skerries, Kilmore Quay, Preston, Mersey

0.5–1.0 Newlyn, River Fal, Brixham, Bridlington, Scarborough, Whitby, Arbroath, Wick, Whitehaven, Milford Haven, Howth, Dunmore

1.0–7.0 North Shields, Eyemouth, Leith, Anstruther, Fraserburgh, Macduff, Buckie, Lossiemouth, Lerwick, Ullapool, Stornoway, Oban, Cambletown, Ayr, Killybegs

7.0–15 Lowestoft, Peterhead, Mallaig, Fleetwood

15–22 Aberdeen

22–30 Grimsby, Hull

This list has not changed significantly in the last decade except that the 'Cod War', which led to the exclusion of vessels from the Icelandic grounds, has resulted in a severe decline of the Hull-based fleets.

Catch by region

The distribution of fishing in the British Seas reflects the combination of oceanographic and biological factors described earlier and the proximity of a suitable fishing port. Table 10.4 illustrates this data by

Table 10.4 Catch by region, 1985

Region	Kilotonnes	%
Barents Sea	0.4	0.1
Norwegian Coast	2.5	0.4
S & W Ireland	3.6	0.5
Faroes	0.4	0.1
West Coast	267.6	39.1
North Sea	349.6	51.1
Rockall	9.0	1.3
Irish Sea	27.9	4.1
English Channel	18.7	2.7
Bristol Channel	3.8	0.6
Others	0.3	0.1
Total	683.9	100.0

showing the landings of sea fish (excluding shellfish) in the UK by British vessels by region of capture during 1985.

This was supplemented by landings from foreign vessels of 355.6 kilotonnes during the same year, giving a total UK supplies weight (excluding shellfish) of 1,029,556 tonnes which had a value of about £500 million. Shellfish landings for the same period were about 100,000 tonnes with a value of £130 million. The total landings rose to 1,120,000 tonnes during 1986 with a value of almost £750 million.

The distribution of fishing

This chapter has been concerned with the utilization of the British Seas as a resource for the fishing industry. As such, fishing is a unique industry, because unlike its land-based counterparts in agriculture, the resource is not actually the property of any individual until it has been taken. In this sense the fishing industry, despite its high capital investment and advanced fish-finding technologies, has not progressed beyond the hunter-gatherer economy. This section attempts to utilize the spatially restrictive analysis which was introduced in Chapter 1 to draw broad conclusions about the distribution of the industry in the British Seas.

Environmental restrictions

The species which have been described in the preceding pages are all distributed across much of the British Seas region; however, it will have become apparent that the oceanographic factors which were detailed in Chapter 6, and which in turn control the phytoplankton production and thus the availability of food, exert an important influence on the distribution of the different species. Cod tends to remain around its spawning grounds, but haddock and whiting extend further and herring and sprat are carried by dominant currents in their larval and young fish stages. The mackerel migrates over relatively large distances with the seasonal sea temperature changes. Again the herring, in particular, spawns on gravel substrates and therefore the environmental potential for this fishery is largely controlled by the processes of seabed sediment transport discussed in Chapter 7.

Thus environmental controls on the distribution of the resource in the British Seas are extremely important, but there is a lot of further work which remains to be carried out on the distribution and population dynamics of the fish before such controls are fully understood (Cushing, 1977).

Economic restrictions

The sea fish industry has a great overcapacity to supply the present market demand. This is because there are too many fishing vessels, and given the problems associated with over-exploitation of many of the fish stocks, it is too restricted by the types of species which it can take. The economic restrictions are not, therefore, those usually associated with commercial ventures, in that the technology does permit almost any species to be recovered from almost any region of the British Seas and to be brought to market at a price which is comparable with other foods. The market will not, however, sustain that price on any but the most popular species, and these are already over-exploited by the present industry.

Policy restrictions

The application of restrictions on commercial fishing is not a new phenomenon. Ashford (1915) comments that, in 1377 during the reign of Richard II, laws were enacted which limited the minimum size fish which could be taken. The law applied to turbot, brill, codling, whiting, plaice, dab, and flounder and to various British rivers, including the Humber, Dee, Thames, Wear, Tees, and Mersey. The minimum sizes ranged from 16 inches (about 40cm) for turbot to 7 inches (about 17cm) for flounders. In 1765 the law was strengthened with the imposition of penalties which included transportation for seven years or imprisonment and flogging. Although the penalties may have eased somewhat, the problem of overfishing has not, and there is a continuing conflict between, on the one hand, the fishing industry which seeks to maximise fish landings and to generate a return on invested capital, and the governments of the European nations who, acting on the advice of their various scientific officers, have sought to regulate the industry in order to prevent overfishing or to retain control on catches in their own territorial waters. The conflict is complex because it is, after all, in the best interest of the industry to preserve the long-term viability of the stock. The debate is also far beyond the scope of this chapter and the interested reader is referred to Shackleton (1986) for a review of the possible policy options which range from intervention, through mediation to a self-help perspective. Alternatively Mirman and Spulber (1982) and May (1984) present interesting attempts at the formulation of qualitative, analytical and numerical models for different policy options from an economic viewpoint. More pragmatically, Farnell and Elles (1984) describe the development of the European Community's Common Fisheries Policy for the British Seas. Their conclusion appears to be valid: 'the argument to date over the shape of the Common Fisheries Policy has obscured the underlying problems of the industry'.

It becomes apparent from these texts that the legislative controls have a severe effect on the industry, and that they range from complete bans (as with herring in the North Sea), through the imposition of maximum net sizes and fish lengths, and through national quotas, to various forms of intervention by way of buying off surplus vessels and grants for capital investment.

References

Ashford, W.A., 1915. Hull as a fishing port. In: *Hull as a Fishing Port*. Publishing Offices, Chapel Lane, Hull, pp. 78–90

Cushing, D.H., 1977. *Science and the Fisheries*, Studies in Biology, 85. Arnold, London, 60 pp.

Farnell, J. and J. Elles, 1984. *In Search of a Common Fisheries Policy*. Gower, Hampshire, England, 213 pp.

King, C.A.M., 1975. *Introduction to Physical and Biological Oceanography*. Arnold, London, 372 pp.

Lee, A.J., 1958. Marine life in relation to the physical-chemical environment. In: Steers, J.A. (Ed.) *Physical Geography*, 4th Edn. Cambridge University Press, Cambridge, pp. 420–455.

Lee, A.J. and J.W.Ramster, 1981. *Atlas of the Seas Around the British Isles*. MAFF, Lowestoft.

May, R.M. (Ed.), 1984. *Exploitation of Marine Communities*. Springer-Verlag, Berlin, 367 pp.

Ministry of Agriculture, Fisheries and Foods, 1988. *Sea Fisheries Statistical Tables: 1986*. HMSO, London, 41 pp.

Mirman, L.J. and D.F. Spulber (Eds), 1982. *Essays in the Economics of Renewable Resources*. North-Holland, Amsterdam, 287 pp.

Sea Fish Industry Authority, 1987. *Annual Report*. SFIA, London, 48 pp.

Shackleton, M., 1984. *The Politics of Fishing in Britain and France*. Gower, Hampshire, England, 407 pp.

Chapter eleven

Seabed mining

The coastal and offshore seabeds of the British Seas are covered with a wide variety of superficial sand and gravel deposits (Chapter 7) which are mined for sand and gravel aggregate, for specific minerals, for carbonates and for navigational purposes. The aggregate industry is the largest in both volume and economic value and continues to grow because the demand for sand and gravel onshore for industrial and commercial purposes has been increasing for centuries. Early uses included dry ballast for sailing ships, but the cost of bulk material transport prohibited the use of anything but local materials except in exceptional circumstances, until the development of rail and inland waterway links in the last 100 years. Thus, the use of marine aggregates was largely confined to coastal areas and the sources were generally within estuaries and along coastlines a short distance from the ultimate destination. The early vessels used in mining inshore deposits were usually conversions based on either fishing boats or small coasters. They were fitted with grabs for both winning the material and discharging on arrival in port. Ships of this type, often no more than 120–150 tonnes capacity, are still operating in estuarial waters, particularly in France.

The use of marine aggregate resources for the construction industry, as a viable alternative to material quarried on land, has recently become increasingly important, particularly as the land reserves became progressively exhausted and at the same time, more difficult and expensive to obtain. Offshore mining techniques have become more refined and the penetration of marine aggregates into the market has increased until, in the 1980s, some 15 million tonnes are being produced each year in the UK. This represents between 15 and 20 per cent of the total UK aggregate production.

The licence to mine seabed material is given by various national bodies in the different European countries. The international rights are defined under an agreement embodied in the Continental Shelf

Plate 11 Aggregate dredger discharging at a riverside wharf (courtesy of
United Marine Aggregates)

Acts by the European governments whose coastlines border the region, and have resulted in the division of the seabed into clearly defined areas. This allocation left the UK with an overwhelming share of the known offshore aggregate reserves. The following account is based largely on Nunny and Chillingworth (1987), with some more specific references in the body of the text.

In addition to the mining of aggregates, seabed sediments can represent a commercial concentration of specific minerals, and deposits of these types (known as placers) have been worked for more than a century in the British Seas. Again carbonate-rich sediments are taken from certain areas for use in a number of industrial processes. Finally the seabed is increasingly dredged to provide safe passage for the growing numbers of deep draught bulk carriers which cross these waters. Each of these four types of seabed mining in the British Seas are described in the present chapter.

Aggregate mining reserves

Seabed sand and gravel is largely the legacy of Quaternary marine transgressions (Chapter 3) which have been reworked by the processes of waves and tides (Chapters 4 and 5) and by modern seabed sediment transport (Chapter 7). We shall see later that dredgers cannot operate economically in depths of more than about 30m and that production licences are rarely awarded in depths of less than 18m. Seabed sand and gravel is not, however, a renewable resource, and therefore the industry is dependent upon continuing to find and to extract aggregate from the strip of coastal waters between these bathymetric limits.

Table 11.1 Estimated known aggregate reserves, 1987

Area	Licensed reserves (million tonnes)	Unlicensed reserves (million tonnes)
South Coast	41	85
Thames Estuary	15	+
East Coast	41	18 +
Humber	32	30
Scotland	None	11
Irish Sea	10 (sand)	12
Bristol Channel	40	5
Total	179	161 +

Note: + refers to additional, undisclosed reserves.

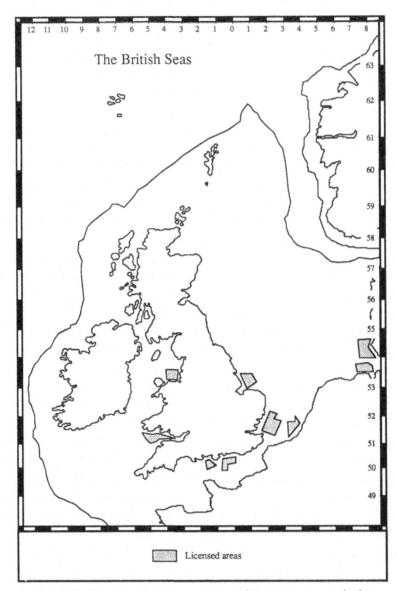

Figure 11.1 Distribution of actively worked marine aggregate areas in the British Seas.

An overview of the present state of aggregate reserves can be obtained from Table 11.1 which utilizes the data presented by Nunny and Chillingworth (1987). This details licensed reserves and unlicensed reserves, where the latter term refers to sites which have been prospected, but on which decisions are either pending or have been refused. The region around the coast of Great Britain is divided into six areas, and each of these will be briefly summarized (Fig. 11.1).

The south coast

This area has never been glaciated (Chapter 3) and consequently the deposits represent periglacial (close to an ice-sheet) and marine reworking of material derived from the nearby chalk and from the Solent river terraces. The area is extensively dredged at present but may also contain additional resources in water which is deeper than the present practical limit of about 25m.

East coast and Thames Estuary

Extending from the coast of Kent to north Norfolk, this area is shallow and represents a complex Quaternary history of river terraces, glaciation, and marine reworking. The Thames valley drainage systems have left a complicated sequence of deposits which extend from the Thames Estuary to north of Norwich, and since the most recent glaciations did not extend so far south, this area evidences marine and periglacial reworking. Further north the last, Devensian, glaciation has left a layer of till, with a thin gravel lag which was deposited during the subsequent sea-level rise. All of these areas are actively dredged.

Humber

Extending from Lincolnshire to Northumberland, this area was repeatedly glaciated, and north of Flamborough the seabed is steep and has been extensively reworked by subsequent marine action, leaving only localized pockets of sand and gravel. Off the Humber itself, local till has been winnowed to produce a lag deposit which thickens and is dredged from the region of the Inner Silver Pit. These sources have probably been augmented by fluvio-glacial material associated with an earlier, seawards mouth of the estuary. The area is dredged by vessels from Hull and from further south.

The resources

Scotland

Scotland's long and intricate coastline probably represents the greatest potential for aggregate reserves, but as yet there has been little commercial work. This may be because the relic gravels lie above rather than below sea-level because of the isostatic emergence of the area since the last glaciation (Eden, 1970).

Irish Sea

The northern Irish Sea and much of the southern areas were glaciated and the present-day seabed represents locally concentrated deposits of glacial and fluvio-glacial sediments, and farther south there are some moraines. The gravel areas tend to be too far away from the Mersey market, or to be in too deep water for exploitation, and instead coarse sand is dredged along the North Wales coast.

Bristol Channel

Partially glaciated, this area represents a complex combination of fluvio-glacial and lag deposits which have been extensively reworked by the intense wave and tidal actions in the region. Aggregate reserves have been identified on banks along the South Wales coast, but the present production is almost exclusively of a coarse, gritty sand from the inner channel.

Aggregate mining technology

The industry, in the last 50 years, has developed from the early, inshore bucket dredgers. The first suction pumps came into regular use after the Second World War, initially for the mining of sand deposits. There has been steady development over the years in both ships and equipment, but it was not until the mid-1950s that purpose built offshore marine aggregate vessels began to appear. These first custom-built ships were in the 600–650 tonne range. The average available pump power and pipe length still restricted them to the exploitation of relatively shallow inshore banks. Through the 1960s larger vessels began to appear both as conversions of suitable small bulk cargo ships and as custom-built vessels.

During this period of expansion, various pump and pipe layouts were used. Forward facing pipes were initially popular but restricted the vessel to static mining. Another variation was the trailing pipe system coming inboard at deck level. This simplified installation but

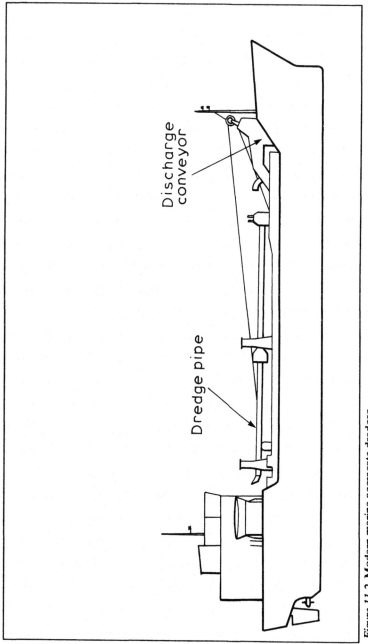

Figure 11.2 Modern marine aggregate dredger

lost out on depth penetration. The system which steadily began to emerge as the best overall combination was the trailing pipe with slide fitting (Fig. 11.2). This ensured that the ship could trail whilst loading, therefore making use of shallower raw material deposits at maximum depth resulting from the extra length gained with the slide fitting in the ship's side.

This layout had various other direct benefits. The internal pump installation operates at greater efficiency at the bottom of the slide. The installation of a slide, avoiding the use of flexible hoses over the ship's side, ensured a much more solid operating position. There is obviously much less pipe movement when the slide is locked into place compared with the flexible overside operation which is very susceptible to adverse sea conditions prevailing between sea surface and deck level. Swell compensation units also became standard equipment, thus enabling vessels to continue operating in sea conditions which had previously caused operations to be suspended. The newer vessels now in service are larger, usually having a loaded capacity in excess of 3,000 tonnes of cargo. They have more efficient mining systems and self-discharging facilities. This latter development has improved turnround in ports dramatically. Whereas the older grab discharge vessels usually remain in port until the next tide (12h), the self-discharge vessels expect to turn round in a maximum of 4 hours, thereby saving the tide and offering a considerable productivity bonus.

The demand for marine aggregates

Commercial viability of a seabed mining operation requires first, as with any product, that there must be a market demand and second that there must be a source of suitable material within economic distance of that demand. Two distinct although interrelated market demands have developed for marine aggregates. On the one hand large volumes of material have been required in short time spans for major construction developments which have direct access to the sea. The requirement is either for bulk structure fills or for sand and gravel for concrete. Examples of this short-term, high-volume usage are the port developments at Rotterdam, Zeebrugge, and on the River Thames and the nuclear power plants in Brittany and at Sizewell in England. The alternative market to such short-term, high-volume usage is the long term supply of aggregates to the building and construction industries. The marine aggregate industry worked hard in the 1960s convincing users – and perhaps itself – that it was a serious long-term source of materials for the construction industry. The introduction of materials from offshore clearly raised problems of acceptability, some of which still exist today.

The 1970s saw a growing acceptance of marine aggregates in lieu of land gravels in the UK construction industry, particularly in southeast England, where more than 60 per cent of total requirements are now serviced from offshore. This increase in demand encouraged considerable investment in offshore equipment and in the 1970s more than thirty companies operated about eighty ships and produced approximately 14 million tonnes. However, by the end of the 1980s twelve companies were operating fifty vessels and producing about 15 million tonnes annually. This rapid change was caused by a series of mergers and takeovers as a small number of the major onshore producers found it desirable to preserve direct access to their own source of marine material as their land deposits were depleted. This trend will continue.

In France at this time, there are twenty-nine companies with forty vessels producing 5 million tonnes annually, in a position which resembles the UK before rationalization.

The supply of marine aggregates

The mining of marine aggregates in the British Seas is illustrated by Fig. 11.1. The UK is extracting sand from the Severn Estuary, and sand and gravel from the Bristol Channel, the Scilly Isles, the Solent and East Wight, the Thames Estuary, the Suffolk and Norfolk coasts, and off Lincolnshire and Humberside.

The Danish industry extracts marine aggregates from the Baltic and from Denmark's northern North Sea coasts, but the volumes have contracted rather than expanded during the 1980s. Material was being supplied to the West German Baltic ports but this trade has now ceased and there is presently little sign of any long-term, large-scale development.

Despite the internationally recognized position of the Dutch dredging industry there is presently only minimal estuarial mining of sand and no regular offshore work. The small Belgian area (Fig. 11.1) is being actively worked for local demand and this is supplemented by British vessels operating from UK licences. The French are the second largest extractors of marine aggregates on the UK shelf and produce about 5 million tonnes annually from around Brittany. The French industry also imports material from the UK and extracts limestone for fertilisers.

A comparison of Fig. 11.1, which shows mining areas, with Fig. 7.4, which shows the distribution of superficial seabed sediments, exemplifies the economics of the offshore aggregate industry. It is clear that potential resources are far more extensive than present-day mining areas. The limit is imposed by three factors:

1. In order to win a licence to produce from any of the national controlling bodies the company must satisfy the objections of the various parties which prevent extraction from many areas including fishing grounds and fish spawning regions, offshore oil and gas rigs and pipelines, submarine cable routes, heavily trafficked shipping lanes, and sensitive military areas.
2. Although modern dredgers are capable of operating efficiently in up to 30m of water, licences are rarely granted in less than 18m because of fears of coastal erosion, so that a ribbon between these bathymetric contours is defined.
3. Transport costs between extraction site and the customer define areas within economic reach of port operating, desalting and grading facilities.

These restrictions have resulted in the distribution of active aggregate mining shown in Fig. 11.1. The following details of each site are based on Banner (1980). Large-scale extraction began in the Bristol Channel in the early 1960s, where well-sorted sand from linear offshore banks found a ready market as a mixture with the quarried stone of South Wales (Hill, 1971). London-based operations began in 1962, with open-sea dredging in the Sunk area off the Thames Estuary, and operations now extend far into the North Sea.

Except in the Bristol Channel, where sand alone is extensively worked, the main target is gravel, and ideally a mixture of 40–60 per cent gravel with clean quartz sand with no more than 5 per cent mud or 3 per cent carbonate content. Coastal gravel deposits, from recent erosion of the coastline, are worked extensively only when it is necessary to dredge them (Hill, 1970), but many coastal gravels of Quaternary origin constitute large, untapped reserves. For example, a single gravel spit in the Moray Firth which developed in late glacial seas but which has been extensively modified by recent marine action extends 1.5kms seawards from the beach and has been estimated to contain 2 million tonnes of gravel reserves (Harris and Peacock, 1969). However, most deposits currently being exploited are those in the sheltered waters of estuaries (e.g. the Harwich banks in Essex); the largest such source is the River Tay which produces about 100,000 tonnes each year. The most rapidly developing area is the southern North Sea where Thames-based dredgers operate to supply London, Kent, and the continent (Fig. 11.1). Liverpool and Manchester are supplied with over 1 million tonnes annually of sand from the Mersey Estuary and gravel from Liverpool Bay. The Bristol Channel supplies about 2 million tonnes annually and this represents about 80 per cent of the aggregate used for construction in South Wales (Thomas, 1973). Newer dredging areas have been opened up more recently off the Humber and the Lincolnshire coast and these are winning more than 2 million tonnes annually, with about half being exported to the continent (Archer, 1973), whilst further south Thames-based dredgers are taking about 4 million tonnes of aggregate annually from the Shipwash, Sunk and Gabbard Banks off the Thames Estuary. The English Channel has been estimated to contain more extensive reserves of retrievable sand and gravel but active commercial exploitation is concentrated off the west and south-west of the Isle of Wight.

Marine placer mining

In addition to the sand and gravel aggregates described in the preceding sections, the British Seas are mined for metal, in the form of placer deposits, and for calcium carbonate, in the form of shell debris. Metalliferous minerals must, at some stage of their exploitation, be extracted from disintegrated mother-rock and then be concentrated by separation from the mother-rock waste. This may form part of the industrial mining process or it may be accomplished naturally by the processes of erosion, transportation, and deposition of the rock on land, in rivers, and in the sea. The stable minerals (gold, magnetite, cassiterite, and so on) may, in particular, be naturally concentrated as *placer* deposits when lighter fractions are hydraulically winnowed away. The eighteenth-century gold rushes in Australia and America, for example, were triggered by the discovery of the precious nuggets in concentrated placer deposits in streams and rivers, and the miner continues this process with the traditional panning of alluvial sediments to wash away the fines and reveal the gold.

In the British Seas, the only economic placer deposit which has so far been exploited commercially is cassiterite, an oxide of tin, which has a specific gravity of 6–7 and may be concentrated as the lighter minerals (for example quartz sand with a specific gravity of only 2.65) are winnowed away during weathering and transportation. The source rock for the placer deposits may be primary and either igneous or metamorphic, or it may be a secondary sedimentary rock which already contains an abnormally high concentration of the mineral. In the present section we seek to identify the origin of the cassiterite placers in the British Seas, and thus to explain the distribution of modern commercial deposits.

197

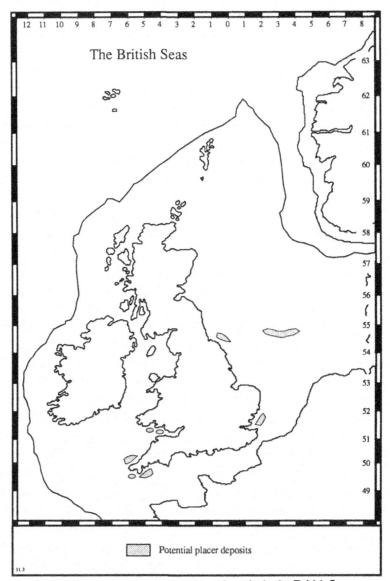

Figure 11.3 Distribution of potential placer deposits in the British Seas.

During the Permo-Carboniferous granites were intruded into what is now western Cornwall (Chapter 3) and it is to these that the area owes the mineral wealth which has been exploited since the Bronze Age. Lodes of copper, tin, and lead were created in the country rock, and are still mined directly around Cambourne, Redruth, and St Just. The waste from tin mining over more than 2,000 years, together with the natural weathering of outcropping lodes through most of the Cenozoic has led to extensive deposition of cassiterite-bearing placer sands in both streams and along the shallow marine foreshore.

There has been active interest in commercial retrieval of this tin from off the coast of Cornwall for over a century and the British Parliament passed a statute in 1853 vesting in the Crown the rights to all Cornish submarine minerals. Methodical exploration of the seabed has taken place only in the last few decades, but this has proven that the offshore deposits are not commercial at present-day prices and in comparison with those remaining on the mainland. The Mounts Bay and St Ives Bay area in south-west Cornwall (Fig. 11.3) has streams which contain up to 20,000ppm tin as placer cassiterite and the mean content (9,000ppm) is 150–300 times greater than in the country rock (Tooms *et al.*, 1965). However, hopes of an offshore deposit cannot be high, because of the extreme wave and tidal conditions which predominate in the exposed region (Chapters 4 and 5). A small treatment plant was established in 1967, and the dredger *Baymead* began working off the Red River Estuary in the same year to retrieve cassiterite concentrated in the top 0.6m of seabed sediments. The dredger separated the sediment, and the denser material was piped back to a sand basin in St Ives Bay, milled and re-sorted at the quayside plant to produce a residue that was about 20 per cent tin. However, downtime due to adverse weather and tidal limitations of the local harbour led to financial losses and the closure of the venture. This emphasized the need to take 'oceanographic downtime' fully into account when assessing probable expenditure. Banner (1980) concludes that the remainder of the British Seas are likely to be commercially unproductive for this resource, because no other area possesses both the source rocks and the hydrodynamic environments conducive to placer cassiterite deposits.

Another placer mineral, zircon ($ZrSiO_4$), which is also relatively dense (specific gravity 4.7) has been shown to be present in the British Seas. Beg (1967) reports pockets of zircon rich sand in the linear banks of the Thames Estuary. Hill and Parker (1971) also report that concentrations as high as 291–300 ppm zircon have been found on the Shingles Patches and on the adjacent west slopes of Long Sand about 5.5km east of Maplin Sands in the southern North Sea. The

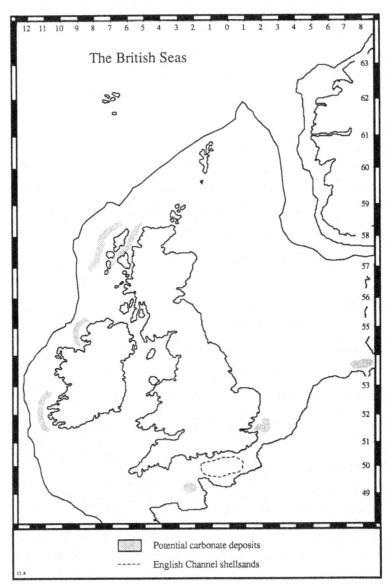

Figure 11.4 Distribution of potential carbonate deposits in the British Seas.
Source: Based upon the data given by Banner (1980).

derivation appears to be the Tertiary sediments of the London Basin. However, even these concentrations are not, of themselves, economic unless dredging of the material was in any case in hand for some other reason. The highest proven concentrations of zircon occur on the westwards-facing beaches of South Wales. Here again ocean swell processes have produced concentrations of around 1,100ppm at Rhossili Bay on the Gower Peninsula and from Abervon to Porthcawl in Swansea Bay. This material is probably derived from geologically earlier concentrations in the Carboniferous millstone grits and coal measures which here form much of the coastal outcrop and also floor the Bristol Channel. The seabed outcrops of tertiary sediments in the north-western North Sea appear to have been the source of local concentrations of zircon on beaches near Spura Head (775ppm), Druridge Bay (1,035ppm), offshore from Alnmouth, Northumberland (430ppm), and on the Dogger Bank (365–1000ppm).

There are a number of other, unproven, reports of placer minerals on the seabed or along the beaches of the British Seas. These include rutile (titanium oxide), ilmenite (iron and titanium oxide), and monzanite (containing thorium, yttrium and europium) in Budle Bay, Northumberland; ilmenite on the north coast of Jersey, and magnetite sands 'in the seas around the Scottish Highlands' (Dunham, 1967). Banner concludes that cassiterite is the only proven commercial offshore placer at the present time, and recalls Slichter's Law that the probability of a prospector's ruin is P^{-NP} where N is the number of ventures and P the probability of success in a single venture.

Marine carbonate mining

When limestones, which are used for cement manufacture and fertilizers, are rare on land, marine carbonate accumulations become valuable as a local resource. This is an example of marine mining in the British Seas which has not developed into a large industrial concern, but for which the potential certainly exists. Banner (1980) reviews the accumulations, and those data are plotted on Fig. 11.4 and described here.

Cement is calcium carbonate ($CaCO_3$) calcined with mud or clay or shale containing alumina, silica, and (often) iron oxides. Cement muds are dredged from Loch Larne and the Thames Estuary (Dunham and Sheppard, 1970) even though, in the latter case, the source is close to the vast chalk deposits which surround the London Basin. Potentially exploitable biogenic carbonate accumulations occur in parts of the Sea of the Minches, the Sea of the Hebrides, and the

Sound of Iona (Cucci, 1977) where debris of coralline alga, barnacles, molluscs, *Lithothamnium*, and echinoderms can reach 70 per cent $CaCO_3$ by volume. Licences have been issued for the limited removal of shell material from beaches in Scotland which may have its origins in these offshore accumulations and has been carried ashore by wave action and tidal processes.

The so called 'coral strands of Connemara' are also *Lithothamnium*, and occur from the beach to depths of as much as 40m offshore, whilst farther south at Mannin and Kilkieran Bay shell debris produces an almost pure carbonate deposit. Mollusc debris and foraminiferal tests can also comprise the bulk of many beach and nearshore deposits on the west coast of Ireland, particularly at Dogs Bay, Connemara and Dovey Beach, Donegal.

On the continental coast, the Netherlands and north-west Germany are particularly short of limestone and the shell deposits at the mouth of the River Jade have been exploited for centuries both for limekiln supplies and for roadmetal. The source of these intertidal deposits appears to be the flats to the north-west of the estuary, which contain a dense population of bivalves (up to 3,000 per m²). Scafer (1972) reports that one company alone dredges 70,000m³ of shell annually from between Borkum Island and the Jade Estuary, to be processed at Vareler Hafen.

Such exploitation will depend upon local demand; the accumulations of shells at, for example, the Loughor Estuary in South Wales are not mined because local supplies of quarried limestone are adequate. Similarly the rich shell sands of the English Channel, although almost wholly $CaCO_3$ in composition, are unable to compete economically with the limestone deposits quarried in adjacent coastal regions. However, the shell debris which comprises 46–60 per cent of the beach deposits of eastern Herm in the Channel Islands and extends in high concentrations into some 20m of water offshore, is exploited for local fertilizers and for tourism.

The dredging industry

As the tonnage and draft of ships using the harbours and anchorages of the British Seas (Chapter 8) get greater, so navigation channels become more hazardous unless continual deepening and widening is undertaken. The marine dredging industry has grown up to serve these demands and operates alongside the aggregate mining activities described above. Unfortunately it is only rarely that dredging recovers material which has any commercial value, and dredging for navigational purposes usually involves the additional problems of disposing of the material at suitable spoil sites.

The problem of maintenance dredging is not a new one, and Gower (1968) provides a detailed history, claiming that 'the art of dredging undoubtedly began in the riverine dwelling communities in the valleys of the Nile, the Euphrates, the Tigris and the Indus'. He cites the commissioning in 1540 by Henry VIII of the first naval dockyard as demanding almost at once marine maintenance dredging of the harbour bar, a position which was exacerbated when the monarch promoted the construction of larger and larger ships, which he armed with batteries of heavy guns, all of which increased the draft and dredging requirements. In 1909 the new Port of London Authority's first task was to dredge a channel from the Nore Light Ship to London Bridge, and the task increases exponentially with the draft and tonnage of ships. Modern bulk carriers of about 300,000 tons draw 22–24m of water, of 400,000 tons draw 27m and the latest of 800,000 tons draw some 30–31m when laden (Ranken, 1971). A study by the Shell oil company at the beginning of the 1970s revealed that no European port then had an entrance channel of much more than 16m at low tide and only the terminals at Heligoland, Bantry Bay, and the Firth of Forth could take the newest supertankers. Banner (1980) concludes that for such ships the whole of the British Seas must be regraded simply as harbour approaches, and as such they will continue to pose a severe hazard to navigation. The argument is illustrated by Fig. 11.5, in which the mean tidal ranges given in Chapter 3 have been used with the bathymetric data given in Chapter 1 to calculate the depth of water at low tide across the shelf, and all areas with less than the draft of the supertankers described above (31m laden) are shown. There is remarkably little water left for manoeuvre.

The need for removal of large quantities of seabed material is evident, and the dredging industry has developed to serve this need. However, dredging such vast volumes is costly. Deepening the Thames Estuary channels by 1m would have cost £750,000 in the late 1960s (Cloet, 1967) and even this would have been greatly increased if consolidated rock had to be removed. Dredging for the new Milford Haven terminal cost more than £6 million for a 14km-long channel.

Maintenance dredging is now common in many coastal areas around the British Seas, though the busier sea-routes present the greatest hazards. The eastern English Channel and the southern North Sea, which combine through routes with major port approaches and a seabed littered with relict moraines, sandbanks, and sand waves comprise one of the most dangerous areas in the world ocean. The *Texaco Caribbean*, *Brandenburg*, and *Niki* all sank on the

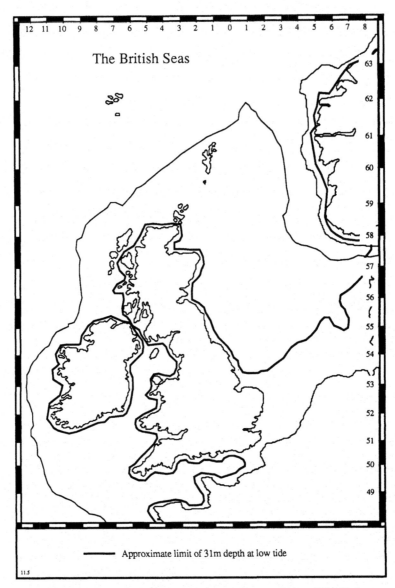

Figure 11.5 Distribution of navigable water at low tide for an 800,000 dwt bulk carrier drawing 31m.

north-west side of the Varne Bank in the English Channel during January 1971, and their wrecks pose additional problems.

The distribution of seabed mining

This chapter has been concerned with the utilization of the British Seas as a seabed mining resource; it appears that the marine aggregate industry is the only commercially viable interest at the present time. This section attempts to utilize the spatially restrictive analysis which was introduced in Chapter 1 to draw broad conclusions about the distribution of such aggregate extraction.

Environmental restrictions

The whole of the British Seas is, in principle, open for utilization by the marine aggregate industry, but in practice the technical problems of deep dredging restrict exploitation roughly to those areas designated in Fig. 11.5. The problem is exacerbated by the complex Quaternary history of the region; in general north of the limits of glaciation the lag deposit must have been sufficiently concentrated to provide commercial reserves, whilst south of this line the localized effects of the drainage patterns and fluvio-glacial and marine action predominate.

Economic restrictions

This book cannot enter into a detailed description of the economic restrictions on aggregate extraction in the British Seas. The reader should refer to more specialized texts such as Nunny and Chillingworth (1987), who consider the following influences:

1. Royalty rates and rent
2. Dredger and dredging costs
3. Wharf and processing coasts
4. Wharf to customer transportation costs
5. Intra-industry pricing structures

Policy restrictions

The two-stage process by which a company must first of all obtain a licence to prospect, and then a licence to produce imposes a necessary but costly policy restriction upon the development of new aggregate reserves. In particular the present-day restriction to water greater than 18m deep prevents extraction from many potentially profitable areas.

References

Archer, A.A., 1973. Sand and gravel demand in the North Sea – Present and future. In: Goldberg, E.D. (Ed.) *North Sea Science*. MIT Press, Cambridge, Mass, pp. 437–449.

Banner, F.T., 1980. Sea-bed resources, potential and actual (excluding hydrocarbons). In: Banner, F.T., Collins, M.B. and Massie, K.S. (Eds) *The North-West European Shelf Seas: the Sea Bed and the Sea in Motion*, Vol. II, *Physical and Chemical Oceanography and Physical Resources*. Elsevier, Amsterdam, pp. 547–567.

Beg, I., 1967. Petrological and environmental aspects of sediments of the Thames Estuary. Unpublished thesis, University of London.

Cloet, R.L., 1967. Determining the dimensions of marine sediment circulation. *Conference on the Technology of the Sea and the Seabed*, Part 3. HMSO, London, pp. 546–563.

Cucci, M., 1977. Biogenic sediments around the Scottish coast. *Reef Newsletter* (Dept. Geol. Univ. Coll., Cardiff), 4, 6–7.

Dunham, K., 1967. Economic geology of the continental shelf around Britain. *Conference on the Technology of the Sea and the Seabed*. HMSO, London, pp. 328–341.

Dunham, K and J.S. Sheppard, 1970. Superficial and solid mineral deposits of the continental shelf around Britain. In: Jones, M.J. (Ed.) *Mining and Petroleum Geology*. Proc. 9th Min. Metall. Congr., London, 1969, 2, 3–25.

Eden, R.A., 1970. Marine gravel prospects in Scottish waters. *Cement, Lime and Gravel*, September, 237–240.

Gower, G.L., 1968. A history of dredging. In: *Dredging*. Proc. Symp. Institute of Civil Engineers. ICE, London.

Harris, A.L. and J.D. Peacock, 1969. Sand and gravel resources of the inner Moray Firth. NERC/IGS, Report 69/6.

Hill, J.C.C., 1970. New legal arrangements needed for marine aggregate extraction. *Hydrospace*, 3, 34–35.

Hill, J.C.C., 1971. Undersea mining of aggregates (2). *Hydrospace*, 6, 30–33.

Hill, P.A. and A. Parker, 1971. Detrital zircon in the Thames Estuary. *Econ. Geol.*, 66, 1072–1075.

Nunny, R.S. and P.C.H. Chillingworth, 1987. *Marine Dredging for Sand and Gravel*. HMSO, London.

Ranken, M.B.F., 1971. Plain sailing in shallow seas is expensive for monster ships. In: Troup, K.D. (Ed.) *Norspec 70: The North Sea Spectrum*. Thos. Reed Publ., London, pp. 39–49.

Schafer, W., 1972. *Ecology and Palaeoecology of the Marine Environment*. Oliver and Boyd, Edinburgh.

Thomas, T.M., 1973. The mineral industry in Wales – A review of production trends, resources and exploitation problems. *Proc. Geol. Assoc.*, 83, 365–384.

Tooms, J.S., 1965. Some aspects of exploration for marine mineral deposits. In: Jones, M.J. (Ed.), *Mining and Petroleum Geology*. Proc. 9th Min. Metall. Congr. 1969, Inst. Min. Metall., London, 2, 285–296.

Tooms, J.S., D. Taylor-Smith, I. Nichoc, P. Ong and J. Wheildon, 1965. Geochemical and geophysical mineral exploration experiments in Mounts Bay, Cornwall. In: Whittand, W.F. and Bradshaw, R. (Eds) *Submarine Geology and Geophysics*. Colston Papers, Vol. 2. Butterworths, London, pp. 285–296.

Wheeler, B.A., 1984. Marine aggregate mining in the UK and Europe: Developments past and present. A.R.C. Marine Ltd, Southampton, pp. 310–319.

Chapter twelve

Wave and tidal power

Energy flows constantly into, and out of the earth's surface environment, and the demands for energy are growing continuously. Fig. 12.1(a) shows that each country's energy consumption increases roughly with that country's gross national product, and Fig. 12.1(b) shows that the rate of growth of energy consumption greatly exceeds the rate of growth of the world's population as a whole. Fossil fuels (coal, oil, and gas) presently provide all but a minute proportion of these energy demands, yet Hubbert (1972) is one of many writers who confirm that, when viewed over the longer span of human history, the fossil fuel era will prove to have been but a short, transitory stage. He suggests that if the period from 3,000BC to 7,000AD is contemplated then the complete cycle of the exploitation of the world's fossil fuels will be seen to encompass perhaps 1,300 years, whilst 80 per cent of total fossil fuels will be extracted and burnt over a period of only 300 years.

What, then, will provide industrial energy in the future on a scale at least as large as the present one? The answer lies in man's growing ability to exploit other sources of energy, chiefly nuclear at present but eventually perhaps the much larger source of solar and gravitational energy. It is the marine aspect of these alternative energy sources which are considered in this chapter. Table 12.1 shows that the earth's surface receives about 173×10^{12} kW of energy from the sun, about 3 per cent of which generates wind waves (Chapter 4) whilst a similar amount of tidal energy is generated by the gravitational attraction of the moon and the sun on the waters of the world oceans (Chapter 5). Serious attempts, particularly in the British Seas, are now being made to harness this renewable energy resource and they are described in the following sections.

Plate 12 Waves, upon reaching the coast, can provide a valuable source of renewable energy (courtesy of Michael Jay Publications)

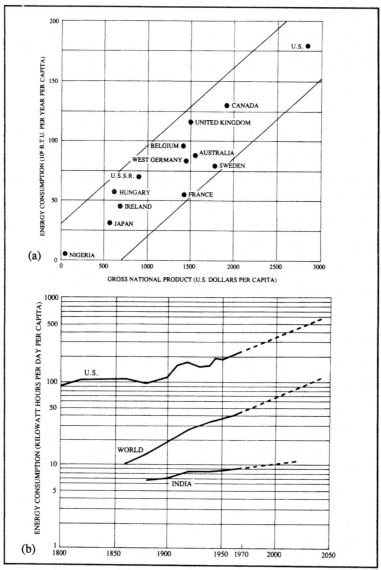

Figure 12.1 (a) Energy consumption against GNP and (b) rate of increase of energy consumption.

The resources

Wave energy

Solar energy, incident on the earth, causes differential heating of the atmosphere which results in circulation patterns which we know as winds. Wind blowing over the surface of the ocean causes waves to be generated by the transfer of energy and momentum from the wind field to the waves (Chapter 4 and Kinsman, 1984). Coastlines on the downwind side of large oceans receive wind energy from the whole of that ocean and waves effectively integrate the wind energy over the ocean and transfer it to the shore. Waves have a higher energy density than wind or direct solar radiation and also have a greater persistence as they do not only depend on local conditions. The highest wind belts occur between 40° and 60° of latitude and hence the western areas of the British Seas, which lie at the downwind margins of large oceans within these latitudes, should experience a large wave energy resource. The power, P, transmitted per m width of coastline by regular waves of height H(m) and period T(s) is given by Lewis (1985):

$$P = 0.98 H^2 T \text{ (kW)}.$$

The wave motion which contains this power consists of surface displacements which can be seen propagating across the ocean, and

Table 12.1 Flow of energy into and out of the earth surface systems

Energy inputs:			
Solar radiation	173,000	$\times 10^{12}$	
Gravitational tidal energy	3	$\times 10^{12}$	
Terrestrial energy	32	$\times 10^{12}$	
Energy outputs:			
Short wave radiation	52,000	$\times 10^{12}$	
Long wave radiation	121,000	$\times 10^{12}$	
Solar radiation transformations:			
Direct reflection (SWR)	52,000	$\times 10^{12}$	(30%)
Conversion to heat	81,000	$\times 10^{12}$	(47%)
Evaporation, precipitation etc.	40,000	$\times 10^{12}$	(23%)
Winds and hence waves etc.	370	$\times 10^{12}$	
Photosynthesis	40	$\times 10^{12}$	
Terrestrial energy transformations:			
Conduction in rocks	32	$\times 10^{12}$	
Convection (volcanoes & springs)	0.3	$\times 10^{12}$	

In addition note that the energy of evaporation and precipitation is stored in water and ice, and that photosynthesis leads to energy storage in plants which is taken up by browsing animals, or the plant decays returning heat to the atmosphere, or the plant decays and contributes to the energy stored within the earth as fossil fuel.

Based on the data of Hubbert (1972).

212

of water particle movements below the surface. Machines have been designed to extract either the potential energy of the surface displacements or the kinetic energy of the particle movements below the water surface. These machines are commonly called wave energy devices or wave energy converters and, although a large number of possible designs have been proposed, practical devices are generally limited to those which respond to variations in the surface elevation of the water during the passage of the wave or to the accompanying pressure changes beneath the wave crest and trough. The devices are designed to translate these surface elevations or pressure changes into electrical energy, and most operate through an intermediate conversion into mechanical energy. Examples of the practical devices are described in the following sections, divided into active devices which move during the passage of the wave and passive devices which remain stationary.

Active wave energy devices

Active devices move in response to wave action and do useful mechanical work against a stable reference frame (Fig. 12.2). This work is converted either directly or indirectly into rotary motion to drive an electrical generator. Diagrams (a) and (b) in Fig. 12.2 illustrate devices which require rigid foundations on the shoreline or on the seabed and would thus involve considerable investment in the installation phase of the project. Alternatively devices (c) and (d) are more simply moored, and are therefore less expensive to install. Diagram (e) illustrates a device in which propellor revolutions provide direct rotary motion, but in any case all these devices require gearing to increase shaft revolutions and to provide the speeds at which generators will most efficiently operate.

The 'Salter Duck' (Fig. 12.2(d)) was one of the earliest systems to show that electricity production from sea waves could be an economic proposition (Salter, 1974). In this device a series of cam-shaped floats which roll in response to waves are mounted on a spine. The shape of the float is such that the wave energy is efficiently converted into the mechanical energy of the duck. The problem lies in then converting this mechanical energy into useful electrical energy, and both hydraulic and gyroscopic systems have been tried in the past decade. A 1:10 scale model consisting of twenty ducks along a 50m spine was tested on Loch Ness and the results are reported by Bellamy (1978). The full-scale machines will have individual ducks with a spine diameter of more than 14m and more than 40m in length each generating 2.2MW of electrical power.

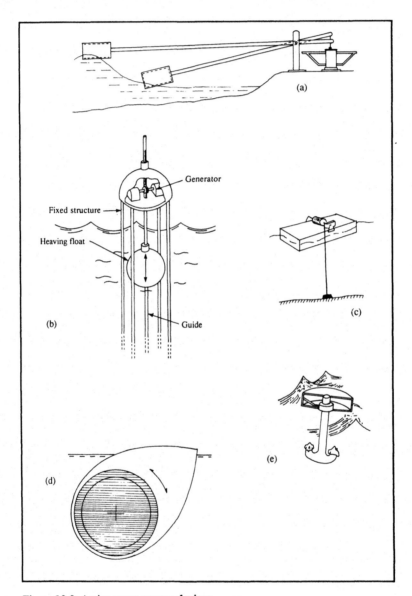

Figure 12.2 Active wave energy devices.

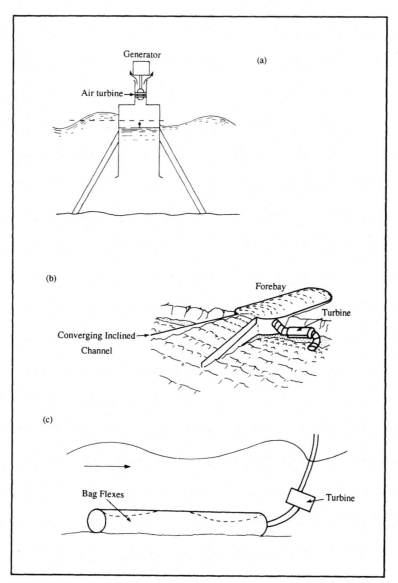

Figure 12.3 Passive wave energy devices.

Passive wave energy devices

Passive devices remain stationary to form the reference frame, and the relative movement of the water is used to perform the work. Surface devices of this type are either the oscillating water column (Fig. 12.3(a)), in which air is forced through a turbine, or overtopping channels (Fig. 12.3(b)), in which the wave energy raises water above the mean sea-level and it returns through a turbine to generate the electricity. Alternatively, air bag devices (Fig. 12.3(c)) are laid along the seabed and pressure variations result in air being driven to a turbine through internal ducting.

The Sea Energy Associates 'SEA Clam' is an economically attractive example of a passive wave energy device which utilizes the principle of seabed pressure variations. It was developed at Lanchester Polytechnic and consists of air-filled flexible bags which breathe air into a central spine in response to wave action (Bellamy, 1982). The spine ducts the pressurized air through self-rectifying Wells turbines to generate electricity. A 1:10 scale model has also been tested on Loch Ness, and the full-scale device is about 290m long and designed to deliver about 10MW. Lewis (1985) reviews a range of wave energy devices and suggests that the SEA Clam could be the most cost-effective for large-scale, offshore deployments.

Wave energy in the British Seas

It has been estimated that the total wave energy resource of the oceans at any one time (Wick and Castel, 1978) is 2TW, (1 terawatt (TW) = 1×10^9kW), which is equal to about one-tenth of the world energy demand for the year 2050 (Isaacs, 1979), although only a small proportion of this can be practically extracted in the foreseeable future.

The estimated total wave energy resource along the European Atlantic and North Sea coastlines is 67.5GW, (1 gigawatt (GW) = 1×10^6kW. This figure represents about 45 per cent of the energy consumption of the European Community. However, these should be treated as gross estimates and measurements of local wave climate (Chapter 4) are essential if a meaningful assessment of the resource is to be made at any specific location. The following detailed assessments are taken from Lewis' review (1985).

The sea areas around the UK have a relatively high density of wave measurement stations (Draper, 1978). Detailed measurements have been made at the South Uist site which reveal an annual mean incident wave power of 48kW m^{-1}. It has been estimated, therefore, that the total UK resource is 30GW (Winter, 1980). There are virtually no wave measurements off the Irish coast, although recent analysis of a station close to the shore suggests an annual mean incident wave power of 55kW m^{-1}, which leads to an estimated 25GW for the whole of that country's coastline. The average French wave energy resource from the North Atlantic is estimated (Ollitraut, 1982) at 16kW m^{-1} with the highest values being recorded in the south and a total of 11GW. The total for the Atlantic coastlines is therefore 66GW.

Wave measurements by the relevant countries along the North Sea coastlines suggest that the annual mean power level is 5kW m^{-1}, although it may be substantially greater at specific localities. Nevertheless these figures suggest that North Sea wave energy resource is only about 1.5GW.

The ultimate measure of the viability of the devices described earlier to convert this wave energy resource into electricity must be the price in pence per kilowatt-hour (kWh). Base-load electricity generation (i.e. power produced by large stations operating at high efficiency) costs about 3–4p per kWh to produce and it is unlikely that wave power can compete with this figure. On the other hand 'peak-lopping' stations which are run only when high electricity demand requires them to do so, generate at costs much higher than the base-load figure. The UK Department of Energy has funded a research programme which has resulted in a significant improvement being made in the efficiency and economics of a number of devices. In December 1978, for example, an initial costing exercise produced estimates of between 30p and 50p per kWh for wave energy devices, yet a year later a similar exercise produced estimates of between 5p and 15p per kWh. Further testing of selected designs is continuing and a small commercial station is presently (1989) being installed on Islay based upon a shore-based oscillating water column chamber and pumped turbine of the type described above. A pilot scheme utilizing the overtopping channel system has also been commissioned on the Norwegian mainland in the northern North Sea.

Tidal power

Harnessing the energy of the tides is a goal which, unlike the wave energy proposals detailed earlier, has been achieved, although only in a few locations even though the potential energy available is very large and is environmentally clean. From the eleventh century, tidal power was used to grind grain, and as late as 1940 a dozen tidal mills were still in operation in Britain. The design of modern tidal power stations, and their suitability to the British Seas is detailed here. The

217

many ways in which arrangements of dams, sluices, and turbines may be operated to produce electrical energy are reviewed by Wilson (1969) and Prandle (1984) but there are, in principle, only two types of practical propositions. These are the single pool scheme, which incorporates a single barrier and the enclosed basin which must drain at low water, and the double pool scheme which is capable of generating energy at any time.

Single pool scheme

The simplest means of generating electrical energy from the tides is to build a single barrier, equipped with sluices and turbines, allow the basin to fill during the rising tide through the sluices, close them at high water and release the impounded water through the turbines at low water. Although it sounds simple, the problems of optimizing the installation and its operation are exceedingly complex, and are dealt with by Wilson and Severn (1972) and Wilson (1980). Although the inclusion of two-way turbines capable of operation during both filling and emptying of the basin may considerably increase the energy generation, it seems that the additional complexity results in installation and running costs which outweigh the additional capacity.

Although the availability of tidal power is predictable, like the tides themselves, well into the future, neither the one-way nor two-way turbines in the single pool scheme overcomes the disadvantage of tidal power generation taking place on a lunar rather than a solar cycle. Additionally, the energy output varies during the neap–spring cycle, when maximum and minimum power outputs are typically in a 3:1 ratio. These drawbacks must be overcome by properly integrating the power station into a national or international electricity supply system. In this way, the problems are solved either by diverting the tidal electricity to a high demand region when the requirements are locally low, or by utilizing excess electricity from the tidal power station itself or from the remaining supplies on the grid to energize a storage system. The most common storage system is again hydraulic, with water being pumped to reservoirs at times of low demand to be reconverted into electricity when demand increases. An alternative solution to the problem lies in the development of two schemes at sites where the tides are out of phase with one-another. In the UK for example, schemes in the Solway Firth and the Bristol Channel would provide complementary power at different times of the day (Chapter 4).

The tidal power station on the **La Rance** Estuary in the Golfe de St Malo is the only tidal power station presently working in the British Seas. The installation also incorporates pump turbines so that

218

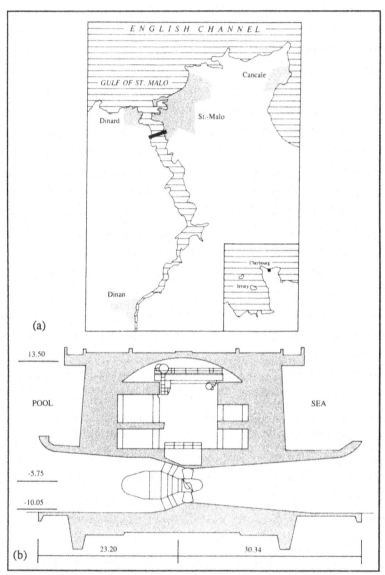

Figure 12.4 (a) Plan of the La Rance tidal power station and (b) cross section through turbines (scale in m).

retiming can be carried out internally. The scheme is shown in Fig. 12.4(a) and a cross-section of the turbines is shown in Fig. 12.4(b). The station was commissioned in 1966 with a design capacity of 544GWh per annum, and the last of the twenty-four 10MW turbines was installed and connected to the network in December 1967. Mechanical and civil engineering details are given by Wilson (1980), and it appears that the scheme is operating extremely well and is generating electricity for some 73 per cent of the time.

Double pool scheme

The classical double pool scheme was proposed by Decoeur (1936) and is capable of generating electricity at any time, or continuously. The turbines are installed between the two pools, one of which is regularly filled from flood tides, the other being regularly drained during the ebb. Pumps may be installed either to pump up the high pool or to pump down the low pool from the sea, or simply to transfer water from the low pool to the high pool through the turbines as at La Rance. Wilson (1980) concludes that although interesting, and often ingenious, these various schemes are likely to remain too expensive, and the problem of retiming the supply will continue to be dealt with by integrating tidal power into a complete electrical supply system.

Tidal power in the British Seas

It has been estimated that the total tidal energy dissipation on the earth is 3×10^9kW (Munk and Macdonald, 1960) and that about one third of this power is being expended in the shallow seas. It is, however, unlikely that more than about 500TWh per annum worldwide is available in tidal ranges of sufficient magnitude or in suitable locations for tidal-electric development. The UK currently uses about 220TWh per annum so that it is apparent that tidal energy can only make a strictly limited contribution to energy supply, although this has been estimated at as much as 10 per cent of demand for the Bristol Channel (Prandle, 1981). Nevertheless, in certain areas, this could be of considerable importance since it can be developed in large amounts and does not use exhaustible resources like the fossil fuels.

Tidal energy may be generated from water turbo-generators if a differential head or level difference can be introduced between two basins, one of which may be the sea itself. Different schemes were outlined earlier, and it has been shown that if the natural tidal range

220

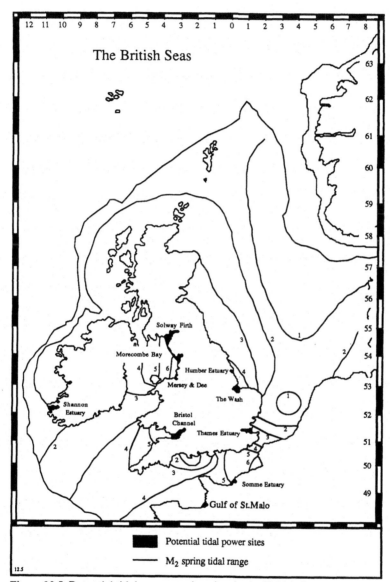

Figure 12.5 Potential tidal power station sites in the British Seas.

The resources

averages less than 5m, then it is unlikely that economic development will be possible. In general (Chapter 5) tidal range increases with distance from the amphidromic points, but local ranges can be amplified by channelling of the coastline and by resonant effects. The distribution of sites in the British Seas with a suitably large tidal range and the possibilities of a relatively short barrier enclosing the generating basin, is shown in Fig. 12.5. There are at least seven sites in England and two in France:

Solway Firth

This site on the west coast of England and Scotland has been investigated by Wilson (1965), and that proposal was estimated to be capable of generating 3.2TWh annually from 100 sets of 16MW generators.

Bristol Channel

The outstanding site for development in Britain is the Severn Estuary. Many proposals have been made for it over the years, and are being considered at the present time. Wilson's (1969) proposal envisaged 120 straight flow turbines each of 9.15m runner diameter and 38MW capacity, producing about 10TWh annually, or about 5.5 per cent of the total electricity sales in the UK. The likely cost of the scheme (at 1976 prices) was estimated at about £1,000 million.

Golfe de St Malo

The scheme at La Rance in the Golfe de St Malo has already been described in some detail and is the only routine tidal power station presently operating in the British Seas.

In addition, there exists the potential for tidal power extraction at the Thames Estuary, the Wash, the Humber, Morecambe Bay, and the estuaries of the Dee and the Mersey in Great Britain and in the Somme Estuary in France.

Wilson (1980) concludes that much remains to be learned; designs need to be refined and improved, hydraulic efficiencies of water passages increased, and the techniques of energy integration and storage perfected, but the principles are established and the practice has been shown to work. It remains to be shown convincingly that large-scale projects can produce energy at competitive rates, and that these can be properly integrated into the energy supply structure to replace fossil fuel sources.

The distribution of wave and tidal power

This chapter has been concerned with the utilization of the British Seas as a marine power resource, for the extraction of both wave and tidal energy. This section attempts to utilize the spatially restrictive analysis which was introduced in Chapter 1 to draw broad conclusions about the present-day and future distribution of such schemes in the British Seas.

Environmental restrictions

The whole of the British Seas is, in principle, open for utilization by the marine power industry, but clearly this is only practical at coastal sites. The preceding descriptions have identified areas of maximum incident wave power to the west of the region, and of maximum tidal range within a suitable basin for tidal energy. These aspects form the major environmental restrictions on the resource, and future developments will clearly be distributed within these areas.

Economic restrictions

The development of either wave or tidal power schemes is extremely expensive in terms of capital investment, and the slow progress at the present time is related to the problems of recouping that investment in comparison to the ongoing use of fossil reserves. This situation will also change in the future as fossil reserves become scarcer and necessarily more expensive, whilst technological developments continue to reduce the cost of marine energy.

Policy Restrictions

At the present time there are few policy restrictions on the extraction of either wave or tidal power from the British Seas, largely because the industry is in its infancy. Any new proposals are, of course, subject to intense scrutiny on economic and environmental grounds, and the proposed schemes in the Bristol Channel and elsewhere are presently being considered through nationally funded impact assessments in this way. Additionally, there is a considerable amount of public money being made available for ongoing research programmes into feasibility studies and technological developments, and these may well prove the potential for the resource in the future.

References

Bellamy, N., 1978. The Loch Ness Trials of the Duck. In: Heathrow Wave Energy Conference, HMSO, London, ISBN 070 580751-7.

Bellamy, N., 1982. Development of S.E.A. Clam Wave energy Convertor. In: *Wave Energy Utilization*. Proc. 2nd Int. Symp., Trondheim, Norway. Tapir Press.

Decoeur, F.,1936. Utilization de l'energie des marees. *Tech.Mod.*, 28, 444–446.

Draper, L., 1978. Marine Information and Advisory Service Bulletin. Institute of Oceanographic Sciences, Wormley, England.

Hubbert, M.K., 1972. The energy resources of the earth. In: *Energy and Power, Scientific American*, Freeman and Co., San Francisco, pp. 31–41.

Isaacs, J.D., 1979. Ideas and some developments of wave power conversion. In: *Wave Energy Utilisation*, Proc. First Int. Symp., Gothenburg, Sweden. Chalmers University.

Kinsman, B., 1984. *Wind Waves*. Dover Publications, New York, 676 pp.

Lewis, T., 1985. *Wave Energy*, Graham and Trotman, London, 137 pp.

Munk, W.H. and G.J.F. Macdonald, 1960. *The Rotation of the Earth: A Geophysical Discussion*. Cambridge Monogr. Mech. Appl. Math., Cambridge University Press, Cambridge.

Ollitraut, M., 1982. Evaluation de la Resource Energetique des Vagues sur la Facade Atlantique Francais. Proc. Conf. on ARGOS Systems, Paris.

Prandle, D., 1981. Tidal power schemes in the Bristol Channel and the Bay of Fundy. *Wave & Tidal Energy*. BHRA, Bedford, England, pp. 397–408.

Prandle, D., 1984. Simple theory for designing tidal power scemes. *Adv. Water Res.*, 7, 21–27.

Salter, S.H., 1974. Wave Power. *Nature,* 249, 720–724.

Wick, G.L., and D. Castel, 1978. The Isaacs Wave Energy Pump – Field tests off the coast of Hawaii. *Ocean Eng.*, 5, 235–242.

Wilson, E.M., 1965. The Solway firth tidal power project. *Water Power*, 17, 431–439.

Wilson, E.M., 1969. Tidal energy development. In: Davis, C.V. and Sorensen, K.E. (Eds) *Handbook of Applied Hydraulics*. McGraw-Hill, New York.

Wilson, E.M., 1980. Tidal power. In: Banner, F.T., Collins, M.B. and Massie, K.S. (Eds) *The North-West European Shelf Seas: the Sea Bed and the Sea in Motion*, Vol. II, *Physical and Chemical Oceanography, and Physical Resources*. Elsevier, Amsterdam, pp. 573–581.

Wilson, E.M., and B. Severn, 1972. Integration of tidal energy into public electricity supply. In: Gray, J.T. and Gashus, O.K. (Eds) *Tidal Power*. Plenum Press, New York.

Winter, A.J., 1980. The UK Wave Energy Resource. *Nature*, 287, 826–829.

Chapter thirteen

Waste disposal

The final British Seas' resource which is considered in this book is the use of the marine environment for the disposal of waste from its highly industrialized coastal countries. Much of this waste disposal is well-organized and well-regulated, and the sheer vastness of the British Seas means that the marine environment is often able to fulfil this role, and to absorb the by-products of society without evidencing any visible effects. This is not always the case, however, and the increasing amount of pollution which is reported in the area is a cause for growing concern.

However, the current emphasis on pollution may create the impression that there has been a relatively sudden deterioration of the marine environment, that was not apparent 20 or 30 years ago. This is not the case. Pollution, in a general sense, must have started at the time when man first began to use the natural resources of the environment for his own benefit. Later, as the human population increased and became concentrated into larger communities, there was an increasing quantity of human and animal waste and rubbish, and sites or techniques had to be found for the disposal of this material. As early as 1273, Edward I made the first anti-pollution law to prevent the use of coal for domestic heating, so smoke pollution at least has been recognized for more than 700 years. By the mid-nineteenth century the population of the UK had risen to 22 million, and the industrial revolution had added industrial waste to the mounting disposal problem. The rivers and canals were by then already grossly polluted, and this effluent, along with the largely untreated sewage from towns and cities, was being discharged into tidal waters. A Royal Commission on the Prevention of River Pollution was established in 1857, and eventually the first preventative river pollution legislation was passed in 1876 and 1890. Even today, however, a number of British and Continental coastal towns discharge almost untreated sewage into near-shore waters.

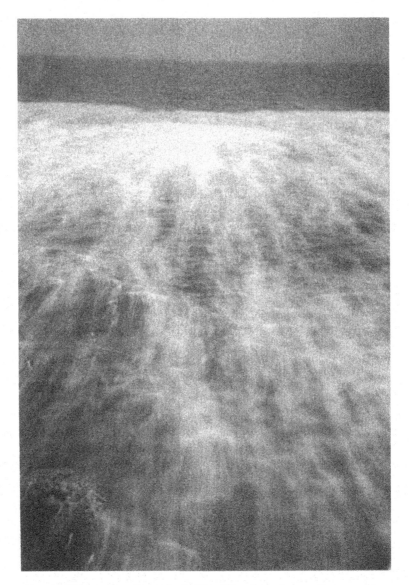

Plate 13 This plankton bloom in the North Sea shows how surface currents
and wind effects can transport material in sea water (courtesy of NERC)

The present chapter is therefore concerned with the present-day distribution of pollutants across the British Seas, and although absolute measurements at sites other than adjacent to coasts are scarce, it is possible to draw some broad conclusions by linking the location and discharges of known inputs with the directions of tidal streams and residual currents which were detailed in earlier chapters. This review is based largely on Dix (1981) and Clark (1986) with other more specific references cited in the body of the text.

The quantification of pollution

Although various definitions of pollution have been proposed, including rather succinctly 'matter in the wrong place', the term may be defined by the phrase, 'a pollutant is a substance or effect which adversely alters the environment by changing the growth rate of species, interferes with the food chain, is toxic or interferes with health, comfort, amenities, or property values of people'. Dix (1981) reviews the subject and finds that pollutants are introduced into the environment in significant amounts in one or more of eight broad sectors, according to the type of producers or industry from which the solid and liquid waste originates:

1. Domestic sector. Domestic premises produce solid and liquid wastes. These range from worn-out furniture, equipment, and cars, gardening waste and refuse, to excreta, grease, washing water, and detergents.
2. Commercial and retail trade sector. These premises produce mainly solids which are restricted to paper, board, and plastics together with obsolete furniture and equipment.
3. Industrial manufacturing sector. This waste is in the form of solids, liquid effluent, and slurries containing a range of organic and inorganic chemicals. Industrial processes are continually changing as new and modified technologies are developed. Consequently products, plant, and premises may become obsolete and worn out, so causing waste disposal and dereliction problems. Also many industrial processes use water for cooling purposes and this can produce thermal pollution if heated cooling water is released into streams, rivers, and lakes.
4. Construction industry sector. Mainly solid waste is produced consisting of brick, stone, mortar, and cement rubble, wood, glass, metals, and plastic as well as obsolete electrical and plumbing equipment and materials. This waste comes from four main types of operations, involving building new

227

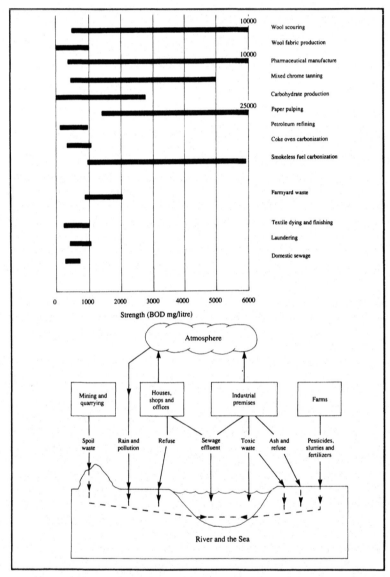

Figure 13.1 (a) The strengths of some major effluents and (b) schematic diagram showing effluent routes into the sea.

premises, adaptation, and modernization of existing premises, the demolition of buildings for land clearance or development and new road construction.

5. Extractive industry sector. This industry carries out various mining and quarrying operations involved in the extraction of coal, rock, slate, sand, aggregates, metallic ores, and clay. The waste consists of unusable solid spoil material, and liquid slurry from washing and grading of the extracted materials.

6. Agricultural sector. Organic wastes are produced from farms in the form of manure slurries, silage effluent, and dairy washings.

7. Food Processing industry sector. This sector includes the production of meat and dairy products, deep-frozen and canned foods, and the processing of liquid and dried food derivatives, ranging from fruit preserves to flour, drinks, and beverages. Wastes range from unusable meat, vegetable, and fruit material, to processing water containing organic chemicals such as fats, proteins, and pesticides.

8. Nuclear industry and power sector. This industrial sector produces cooling water, and solid effluent and slurry wastes that are radioactive for periods of time ranging from a few days to thousands of years.

Each of these will contribute directly or indirectly to pollution in the marine environment because all waste is drained by water which ultimately follows the hydrological cycle into the marine environment (Fig. 13.1). In addition some are specifically dumped at sea and these will produce high local concentrations of pollution. The three most important routes by which waste enters the British Seas are the input of waste into the sewerage systems which drain into fluvial and marine systems, the direct pollution of the seas by the hydrocarbon industry through accidental or intentional spillages and the discharge of radioactive material from nuclear power stations and reprocessing plants. Each of these will be dealt with in the following sections.

It is extremely difficult to assess the potential environmental damage which can be caused by any given pollutant, because the contents may be complex, the processes may be additive and long-term, or seemingly inert and safe material may become highly toxic when mixed with other compounds after disposal. However, the biological oxygen demand (BOD) has frequently been used as an indicator of the potential damage which can be caused. The British Government conducted a water quality survey in 1958 and divided the results into four classes:

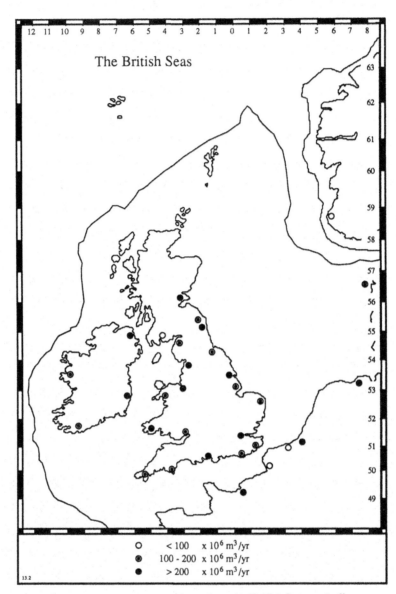

Figure 13.2 Coastal sewage waste input into the British Seas excluding dumping.

Source: Lee and Ramster (1981).

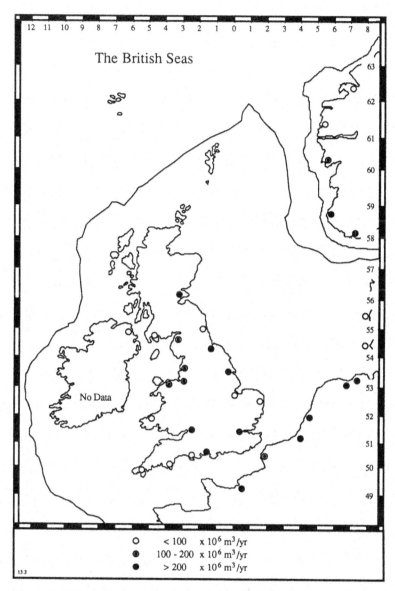

Figure 13.3 Coastal industrial waste input into the British Seas excluding dumping.

Source: Lee and Ramster (1981).

Class 1.   Unpolluted with a BOD below 3mg litre^{-1}
Class 2.   Doubtful condition requiring treatment
Class 3.   Poor condition urgently requiring treatment
Class 4.   Grossly polluted with a BOD of 12mg litre^{-1} or more.

Sewage waste

Sewage is literally the contents of sewers, and these comprise the sewerage systems which carry the water-borne wastes of a community. Sewage originates from domestic and commercial premises, land drains, some industrial plants, and agricultural sites. Industrial waste is legally termed trade waste and is defined in the Control of Pollution Act (1974) as 'any liquid, either with or without particles in suspension, which is discharged from premises used for carrying out any trade or industry, other than surface water and domestic sewage'.

The largest volume of discharged sewage is in the form of effluent which can be any liquid, solid, or gaseous product in a treated or untreated form. The effluent may vary in quality and strength from relatively harmless dirty water to highly toxic sludges and slurries. An idea of the more important types of effluent and their pollution strength in terms of BOD is given by Fig. 13.1(a). This shows that paper pulp and pharmaceutical manufacture, coal carbonization, wool scouring, and tanning all produce effluents with a very high BOD value. Others, including silage and certain chemical processes can produce still higher values of up to 50,000mg litre^{-1} BOD. Efforts are, of course, made to treat these pollutants before they are released into the marine environment either directly, or through the land and river drainage systems, as is shown schematically by Fig. 13.1(b). However, the success or otherwise of these efforts is presently a matter of some conjecture, and it is certain that pollution levels in the British Seas are increasing rapidly, and in some cases have already become critical.

An estimate of the inputs of sewage waste can be gained from Figs 13.2 and 13.3 which show the results obtained from a questionnaire issued by ICES, covering the period 1974–5 (ICES, 1978). It should be noted that the data are of variable quality for they depend upon the ability of the respondents in each country to locate and extract the information. Furthermore, large quantities of sewage effluent pass into the open sea through Europe's rivers, yet the amounts involved have only been quantified for the Rivers Thames and Elbe with 1,027 and 1,278 million m^3 per year respectively. Large but unrecorded inputs from the Rivers Seine, Rhine-Schelde, and Ems are also likely to be important.

232

Oil pollution

The pollution of the British Seas by oil, whether through accidental discharge or deliberate pumping, is becoming a serious problem because of the increasing oil traffic (Chapter 8), and because even quite small spillages can have a catastrophic effect on the local environment. With the increasing numbers of tankers on the sea more accidents have occurred, and as tankers have become larger, the larger vessels have also been involved in more serious incidents. The busy sea-lanes in the British Seas have not avoided the consequences of these disasters. The first major tanker accident was the *Torrey Canyon* which ran onto the Seven Stones Rocks off Cornwall in 1967. The vessel was a complete wreck and 100,000 tons of crude oil were lost. Twelve years later the *Amoco Cadiz* grounded off Brittany in March 1979 and lost its cargo of 220,000 tons of crude oil.

Pollution can also occur as a result of spillage during discharge. The French tanker *Betelgeuse* exploded whilst alongside the Gulf Oil terminal at Bantry Bay in Eire (Chapter 8) with the loss of fifty-one lives in January 1979, and in July 1980 the 212,000 ton tanker *Energy Concentration* broke its back while discharging at Rotterdam. It has been suggested that the structure of these large tankers had been so weakened by corrosion that the extra strain put on them during loading and discharging may have been enough to fracture the hull (Wardley-Smith, 1983).

It is difficult to draw any specific conclusions about the distribution of oil pollution in the British Seas at any one time because there will be a small but variable input from illegal pumping and minor spillages, and these inputs are totally overshadowed by the irregular major catastrophe. However, it is likely that the average distribution follows the shipping lanes detailed in Chapter 8, and that there is a probability of increased pollution wherever these lanes approach navigational hazards, such as the entrance to the English Channel, and in the vicinity of the major terminals.

Radioactive waste

During the generation of electricity at nuclear power stations, irradiated nuclear fuel is produced that has to be reprocessed. In the course of this reprocessing, liquid effluents are formed that are released into the British Seas under strictly controlled conditions. One of the most important components of these liquid effluents is the radionuclide caesium-137, because, together with caesium-134 it has been responsible for the highest level of radiation exposure for the

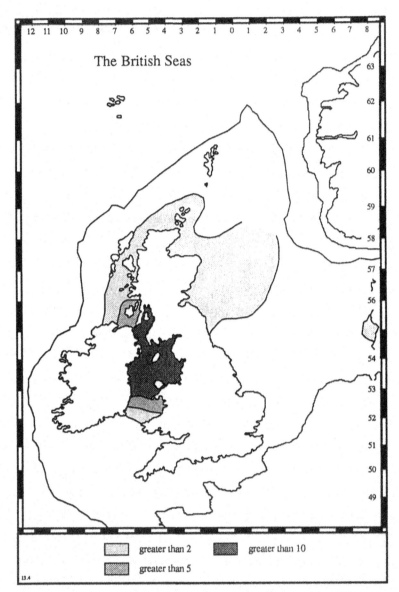

Figure 13.4 The distribution of the radionuclide caesium-137 in 1973.

Source: Lee and Ramster (1981).

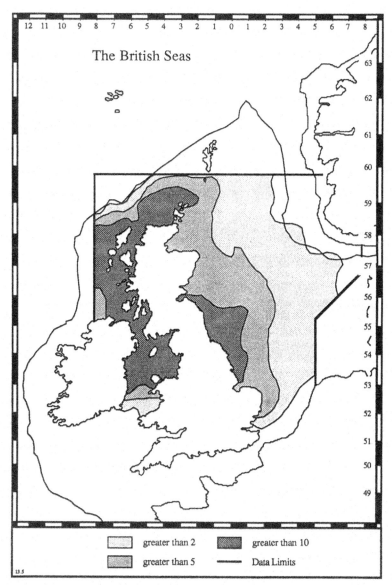

Figure 13.5 The distribution of the radionuclide caesium-137 in 1978.
Source: Lee and Ramster (1981).

235

public at large (Lee and Ramster, 1981). The three main points of
input are the British Nuclear Fuels Limited (BNFL) reprocessing
plant at Sellafield, Cumbria which discharges into the north-east
Irish Sea, the French COGEMA plant at Cap de la Hague which discharges into the English Channel near Cherbourg, and the UK
Atomic Energy Authority's (UKAEA) reactor site at Dounreay, Caithness, which discharges into the coastal waters of north-east Scotland.

For many years the discharges into UK coastal waters have been
the subject of both monitoring and research by the Ministry of Agriculture Fisheries and Food (MAFF) Fisheries Radiobiological
Laboratory (FRL) at Lowestoft, and this work is described in their
technical bulletins. The work assessed the importance of radio-
caesium in human exposure terms, but also utilized the labelling of
sea-water by caesium-137 as a tracer for the examination of water
mass transport (Kautsky, 1973).

The results of these studies are shown in Figs 13.4 and 13.5 which
compare the isolines for the concentration of radioactivity in
pCi litre^{-1} (picocuries per litre, i.e. 10^{-12} curies per litre) in 1973 with
the situation some 5 years later in 1978. It can be seen that the
caesium discharged from Sellafield mixes with the sea-water transported through the Irish Sea in a northerly direction, so that it leaves
the area through North Channel and then moves in a well defined
path northwards along the west coast of Scotland and enters the
North Sea from the north-west. From there it joins the long-term,
generally anti-clockwise circulation patterns which tend to follow the
contours in the North Sea. It is therefore found at varying concentrations throughout much of the North Sea and can still be detected in
the water flowing into the Baltic off the coast of Norway. Work by the
Deutsches Hydrographisches Institut has shown that almost all of
the caesium-137 discharged into the English Channel near Cherbourg is transported first into the southern North Sea and then along
the continental coasts towards the Baltic.

Comparison of Figs 13.4 and 13.5 shows how the concentration
pattern has developed. Prior to 1969, all of the caesium present was
due to fall out from nuclear tests in the 1950s and 1960s, but the three
local sources have clearly had a significant effect since that time
(Kautsky, 1976; Hunt, 1979).

The North Sea: a case study

Despite the environmental stress caused by pollutants throughout
the British Seas, very little is known about their distribution in the
more remote areas to the west, or in the English Channel or the Cel-

tic and Irish Seas. Attention has, however, been usefully focused on the North Sea, due in part to the major industrial inputs through the rivers of eastern Britain and Europe, and in part to the efforts of international groups associated with the European Community. In particular, conferences were held in Bremen, Germany in 1984 and in London, England in 1987 which attempted to at least assess the scale of the pollution problem. In preparation for the second of these, an international Scientific and Technical Working Group (STWG) was established to report on the quality status of the North Sea. Their work is published in STWG (1987), and forms the basis for this section. The STWG took information from a wide variety of sources and considered first the inputs into the region, and second the resulting distribution of contaminants. Each is reviewed separately here.

North Sea inputs

STWG (1987) comments that a great deal of information on inorganic inputs is now available, but that comparison between data provided by different countries is difficult because the data were collected for different individual reasons. However, the results are summarized in Table 13.1, which represents the maximum amounts of input of various metals on an annual basis.

The report also found that the input of hydrocarbons (particularly oil) into the North Sea amounted to almost 30,000 tonnes in 1985. The total included accidental spillages (2 per cent), oil in wastewater (6–9 per cent), but most originated from diesel drilling lubricants (Chapter 9). The use of oil-based drilling lubricants was banned, however, from 1 January 1987, and this input therefore should have decreased. The data are tabulated in Table 13.2, and although the estimates vary widely, it is apparent that, having attempted to deal with the input from offshore hydrocarbon facilities, attention must be directed to rivers and land run-off.

Table 13.1 North Sea inputs in tonnes per year

Source	N_2	P	Cd	Hg	Cu	Pb	Zn
Rivers	1×10^6	8×10^4	52	21	1,330	980	7,370
Direct	1×10^4	3×10^4	20	5	315	170	1,170
Air input	4×10^5		240	30	1,600	7,400	11,000
Dredging			20	17	1,000	2,000	8,000
Sewage	1×10^4	2,800	3	0.6	100	100	220
Industrial			0.3	0.2	160	200	450

Source: STWG (1987).

Table 13.2 Total oil input to the North Sea in kilotonnes per year

Source	Input
Natural seeps	1
Atmospheric	7–15
Rivers, land run-off	16–46
Coastal sewage	3–15
Coastal refineries	4
Oil terminals	1
Other coastal industries	5–15
Offshore oil & gas facilities	29
Sewage sludge	1–10
Dumped industrial waste	1–2
Dredged spoils	2–10
Operational ship discharges	1–2
Accidental/Illegal ship discharge	unknown
Total	71–150

Source: STWG (1987).

The report also discusses the radionuclide inputs into the North Sea, and confirms that, as discussed earlier, the major source is the reprocessing plants at Sellafield on the west coast of England and at Cap de la Hague. Other sources are also identified, and these are listed in Table 13.3. The table has been compiled from STWG (1987) data by summating the total activity for each source. This includes the caesium-137 discussed earlier, along with tritium, strontium-90, caesium-134 and zinc-65 and some other, unidentified materials.

The Cap de la Hague figures in the table are higher than the Sellafield figures because of the greater quantities of radionuclides other than caesium-137 which are input from the French plant. These inputs do not, however, register on Fig. 13.4 and 13.5 because the French plant produces only minimal quantities of caesium-137.

The information contained in the STWG (1987) report represents a best estimate of inputs into one region of the British Seas, and could form a basis for further work in other regions. However, even this information does not adequately describe the present distribution of pollutants, and the report stresses that insufficient data are available to provide reliable estimates of such patterns. Instead, the report argues cogently for the development of some of the modelling techniques which are discussed in the following chapter, in order to provide a sound basis for the control of waste disposal in the marine environment.

Table 13.3 Inputs of radioactivity in liquid wastes to the North Sea for 1984 and 1985

Establishment	Input TBq	
	1984	*1985*
Direct inputs		
Hartlepool ps (UK)	19	22
Sizewell ps (UK)	2	11
Grevelines ps (France)	80	95
Inputs via rivers		
Aldermaston re (UK, Thames)	0.1	0.2
Harwell re (UK, Thames)	2.5	3
Amersham ip (UK, Thames)	0.5	0.3
Bradwell ps (UK, Blackwater)	8	1.3
Chooz Fassenheim ps (France, Meuse/Rhine)	–	30
Doel ps (Belgium, Scheldt)	38	–
Dodewaard ps (Holland, Waal)	0.1	0.2
Borssele ps (Holland, W.Scheldt)	4.6	5.7
Karlsruhe rp (Germany, Rhine)	102	96
Obrigheim ps (Germany, Neckar)	5.0	5.3
Wurgassen ps (Germany, Weser)	0.8	0.7
Stade ps (Germany, Elbe)	12	6
Biblis 'A' and 'B' ps (Germany, Rhine)	32	33
Neckarwestheim ps (Germany, Neckar)	11	13
Brunsbuttel ps (Germany, Elbe)	2.6	0.9
Unterweser ps (Germany, Weser)	25	27
Phillipsburg ps (Germany, Rhine)	2	1
Grohnde ps (Germany, Weser)	0.1	7.2
Krummel ps (Germany, Elbe)	0.6	0.8
Northern inflow of Atlantic water		
Dounreay rp (UK, Pentland Firth)	66	29
Sellafield rp (UK, Irish Sea)*	1,580	2,205
Southern inflow from English Channel		
Cap de la Hague rp (France, English Channel)	2,760	3,867
Paluel ps (France, English Channel)	6	31
Flamanville ps (France, English Channel)	0	0.1
Dungeness ps (UK, English Channel)	7	49
Winfrith ps (UK, English Channel)	113	70
Eastern inflow from Baltic Sea		
Ringhals ps (Sweden, Kattegat)	46	37
Barseback ps (Sweden, The Sound)	1.1	0.6

Key: ps: power station; re: research establishment; rp: re-processing plant; –: no data; *: figures for 1982 and 1983; ip: isotope production
Source: STWG (1987).

239

The resources

Distribution of waste disposal

This chapter has been concerned with the utilization of the British Seas for the disposal of waste materials, and with the results of accidental oil spillages. This section attempts to utilize the spatially restrictive analysis which was introduced in Chapter 1 to draw broad conclusions about the distribution of the waste in the British Seas.

Environmental restrictions

The British Seas are vast. The surface area is almost 1 million km^2, and the mean depth is about 100 metres, giving a total volume of about 10^{14} m^3. It is still extremely difficult to estimate the total volume of pollutant which is released annually into this water body, and the toxicity of the pollutant does, in any case, depend upon its concentration, and on the kind of complex biochemical reactions and interactions which were discussed earlier. However, in order to proceed, consider the data shown in Figs 13.2 and 13.3. It appears that there are about twenty sources each producing in excess of 200×10^6 m^3 of pollutant each year, a total of around 10^9 m^3 per year. This type of arithmetic suggests that, were the pollutants equally distributed, they would be present in concentrations of about 0.01 cm^3 per litre.

Although this figure, in itself, may appear to be small, and one can argue that the waters of the British Seas are continually being interchanged with those of the North Atlantic (Chapter 6), there are two environmental factors which council caution. First, although many pollutants suffer biological degradation under the action of bacteria (and there are even certain bacteria which ingest oils), many pollutants, such as the radionuclides discussed above, are additive and take an extremely long time to degrade naturally. Second, pollutants are not, and never can be equally distributed throughout the water body because the dispersive processes are far too slow. In fact the pollutant is transported from the coastal site of release by the waves, tides and residual currents which were detailed in earlier chapters, and may well accumulate in the same or other areas of the region. The data are not presently available to allow the accurate prediction of these transport paths but, for example, it is clear that the concentration of radionuclides in the North Sea will continue to increase because of the residual circulation patterns around the north of Scotland, as was shown by Fig. 13.4.

Initial work suggests that, in the North Sea, the anticlockwise circulation patterns may be depositing suspended material in the southern Norwegian Channel and off the Skagerrak, although additional sinks in estuaries, fjords, and perhaps in the Wash may be important. The distribution of waste disposal, therefore, bears little or no relation to the eventual destinations of this material, the intermediate paths are presently but poorly understood, and it is obvious that proper control will require a more sophisticated understanding of the environmental processes. This point is discussed further in the following chapter.

Economic restrictions

Proximity to the shoreline and a reduction of processing both lead to more economic waste disposal in the short-term. However, the move towards industrial culpability, in which the body responsible for the release of the pollutant must fund any subsequent cleaning operations, means that there may be a longer term price to pay. It is likely then that waste disposal will continue to be predominantly located in coastal waters, either directly or through river and estuarine inputs.

Policy restrictions

The introduction and implementation of legislative controls on the processing and release of waste materials into the British Seas is likely to become the dominating influence on the industry in future decades. Environmental pressure groups and the emergence of the so-called 'green vote' are persuading governments to take a more aggressive stance, and to license and restrict the release of pollutants. The proper formulation of this policy, however, depends upon a more complete understanding of the dispersion of released material than is presently available. It is therefore likely that policy based simply on restriction will be pursued, whereas an increased utilization of the vast British Seas for the disposal of waste could be contemplated if it were based upon a more correct appreciation of the oceanographic and biochemical processes which are involved. This is discussed further in the following chapter.

References

Clark, R.B., 1986. *Marine Pollution*. Clarendon Press, Oxford.

Dix, H.M., 1981. *Environmental Pollution; Atmosphere, Land, Water and Noise*. Wiley, Chichester, 286 pp.

Hunt, G.J., 1979. *Radioactivity in Surface and Coastal Waters of the British Isles, 1977.* MAFF Direct Fisheries Res., Lowestoft, 3, 36 pp.

ICES, 1978. Input of pollutants to the Oslo Commission Area. Coop. Res. Rep. Int. Council Explor. Sea (77), 57 pp.

Kautsky, H., 1973. The distribution of the radionucleide caesium-137 as an indicator for North Sea mass transport. Dtsch. Hydrogr. Z., 26, 241–46.

Kautsky, H., 1976. The caesium-137 content in the waters of the North Sea during the years 1969–75. *Dtsch. Hydrogr. Z.*, 29, 217–221.

Lee, A.J. and J.W. Ramster, 1981. *Atlas of the Seas Around the British Isles*. MAFF, Lowestoft.

STWG, 1987. *Quality Status of the North Sea*. Report of the Scientific and Technical Working Group, 2nd Int. Conf. on the Protection of the North Sea. HMSO, London, 88 pp.

Wardley-Smith, J., 1983. *The Control of Oil Pollution*. Graham and Trotman, London, 285 pp.

Chapter fourteen

British Seas: a model future

The preceding chapters have presented an introduction to the oceanography and the resources of the British Seas. It will have become apparent that we understand this particular 942,000km^2 of the world ocean at only the most general level, and yet the exploitation of its resources demands careful management if we are not to suffer increased environmental degradation or resource loss. It is appropriate, therefore, to conclude this book in the same way, and for the same reasons, that the lecture course is concluded, with a consideration of the subject matter in the broader framework of environmental research and the future development of the region.

It has become the practice in environmental research to consider three stages in the development of a problem. Alliteratively these are classified as monitoring, modelling, and management; the first refers to the gathering of the basic data, the second to the formulation and testing of hypotheses in order to obtain a scientific understanding of the problem, and finally the scientific understanding is handed over to those who either directly exploit, or to those who must police the exploitation of the natural resources. The following sections deal separately with each of these three stages in relation to the British Seas.

Monitoring the British Seas

It is useful to review Part I of this book through a consideration of the existing data upon which any subsequent modelling is to be based. The availability of data is summarized in Table 14.1, which lists the six areas covered by the individual chapters. The table is divided into three columns, the first of which presents a summary of existing data in each field. In the second column an attempt has been made to emphasize existing synoptic presentations of the data in the form of charts or maps of the British Seas. It is clear that for many of

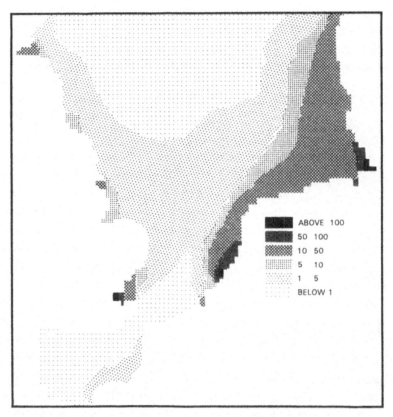

Plate 14 North Sea concentration model detailed in the text (Crown copyright – reproduced with the permission of the controller of Her Majesty's Stationery Office)

the subjects, such synopses are already available, even if at the present time the spatial resolution is limited to large-scale averages and trends. This point will be further discussed below.

Modelling the British Seas

Traditional scientific research proceeds through the erection of hypotheses which purport to predict and to explain relationships between observed parameters. Nowadays, the principle has remained unaltered, but the terminology has changed. It will have become apparent that the British Seas represent a vast and complex series of interrelated processes; however, it should also have become apparent that they appear to obey the fundamental laws of physics. Using supercomputers, mathematical formulations of these laws can be used to simulate events in the ocean and to predict how the ocean will respond to future changes. These simulations, when properly formulated, are in fact hypotheses, but they are referred to as numerical models. To relate so many variables in time and space can only be done with modern modelling techniques based on so-called finite element analyses. The techniques involve the division of the ocean into a large array of cells. The model begins with specified inputs, say a known distribution of salinity, temperature, and the inclusion of certain currents and surface winds. The model then computes the movement of water, energy, and properties such as salt content, or seabed sediment transport. The process is iterated, or repeated a number of times to predict the distribution of the properties at a certain future date.

Some, simplified models of the British Seas have already been constructed and are running to predict the wave climate (Chapter 4), or the tidal movements (Chapter 5). The third column in Table 14.2 attempts to summarize the current state of the art for different models of the British Seas.

However, as the requirement for more accurate and more detailed predictions increases, so the number of cells, the number of iterations and the number of processes in the model must increase. It is likely that computers with this power will not exist until the next century, but scientists are working on providing the basic datasets at the present time. In the late 1980s and early 1990s the British Natural Environment Research Council funded the 'North Sea Project' to develop a simplified model of the North Sea, noting that 'the health of the North Sea in the face of conflicting uses, greatly concerns all the coastal states. Its careful management requires a predictive model of its water quality'. The project built on theoretical analyses

The resources

Table 14.1 The oceanography of the British Seas

	Available data
Shape of the shelf	Bathymetric surveys offer a complete coverage to ± 0.1m in shallow water and with less accuracy over the remainder of the region. Major features have been mapped and named, distant areas along the shelf edge and in deep-water troughs are less well-known.
Geological history	Seismic surveys and bore holes have revealed the major structural and stratigraphic elements of the solid geology throughout the region. Local, detailed results are available or can be extrapolated from land surveys close to the shore or around bore holes.
Wave regime	Long-term records available from wave recorders at about 40 coastal and offshore sites (Fig. 4.2).
Tidal regime	Long-term (permanent) records from tide gauges are available for more than 50 coastal sites and limited-duration deployments provide data at a lesser number of offshore stations (Fig. 5.2)
Oceanographic regime	Temperature and salinity measurements range from a small number (<100) of long-term stations to a large number of unequally distributed instantaneous measurements (>10,000). Similar availability of current meter data.
Modern seabed sediments	Wide coverage of seabed samples (e.g. 1:1,000,000 surface sediment charts), of decreasing density offshore. Southern and eastern seas well-covered by side-scan, less so to north-west and north of region.

Synoptic description	Numerical modelling
Bathymetric charts (Fig 1.1 etc)	Not applicable
Solid geology maps	Some qualitative facies modelling of individual basins (e.g. Ziegler, 1982). Palaeo-climate models (e.g. Henderson-Sellers and McGuffie, 1987) and depositional models (e.g. Pantin and Evans, 1984)
50-year H_{max} charts (Fig. 4.6)	Wave estimates based upon wind stress from meteorological model (e.g. Benwell *et al.*, 1971) could be operated in forecasting mode.
M_2 co-tidal & co-range charts e.g. Fig. 5.4	Davies (1983) with resolution of 1/3° latitude and 1/2° longitude.
°C and ‰ chart (Figs 6.2–6.5) Residual flows (Fig. 6.6)	Davies (1983) above for whole shelf, or e.g. Heaps (1973) for Irish Sea and Heaps and Jones (1983) for Celtic Sea, etc.
Superficial sediments chart	Local models (e.g. Huthnance, 1982, for the Celtic Sea) or shear stress analogues of the transport paths (e.g. Pingree and Griffiths, 1979, cf. Chapter 7).

The resources

Table 14.2 The resources of the British Seas

	Environmental Restriction
Trade and shipping	Port development and navigation requires adequate water depth. Tidal regime controls movement in shallow water, and on steaming with tidal streams. Extreme wave conditions are disadvantageous. Submarine cable routes influenced by nature and stability of seabed.
Hydrocarbons	Distribution of exploitation controlled by geological structure, requiring source, reservoir, and cap rocks. Oil in central and northern North Sea, gas further south. Other smaller fields to the north and west.
Fishing	Distribution of spawning and feeding grounds controlled by nature of seabed and by productivity of seawater, and hence by temperature, salinity, light levels, and stratification.
Seabed mining	Distribution of sand and gravel aggregates largely controlled by Quaternary glaciations and fluvio-glacial environments at time of lowered sea-level, although modern processes are redistributing the material.
Wave and tidal power	Largest waves along western boundary of region. Maximum tidal ranges are distant from amphidromic points, but power extraction also requires the presence of a large, natural basin.
Waste disposal	British Seas are vast, but coastlines are heavily industrialized. Environmental potential for waste disposal is presently poorly understood, but it is related to residual circulation patterns.

Economic & Policy restrictions	Model development
Supply and demand through freight rates. Restrictions through certain shipping clear lanes.	None of which the author is presently aware.
Exploitation limited to larger, shallower water fields by international oil prices. Policy restrictions imposed through the exploration and production licensing procedures, and through taxation.	The industry has finite element oil and gas field models to determine optimum operating conditions (e.g. the British Gas *PROGRESS* model).
Controlled by market price, with severe restrictions on quotas, minimum fish sizes, and the banning of certain species because of overfishing problems.	Some, simple population dynamics modelling (e.g. May, 1984).
High-volume, low-value product which is highly sensitive to transportation costs. Policy controls imposed through the exploration and production licensing procedure restricts dredging to about 18–35m depths.	None of which the author is presently aware.
Only one site operational. High initial investment. Restrictions are re-timing of power availability, and those of public enquiry into future developments.	Economic analyses are linked to the wave and tidal models listed in Table 14.1 (e.g. Prandle, 1981)
Economics demands proximity to source of waste, hence largely river and coastal disposal. Currently extreme political pressure on industry to limit and cleanse waste products.	Some coarse-grid, large-scale models of pollutant dispersal exist (e.g. Plate XIV and STWG, 1987)

of tidal mixing at fronts (Chapter 6), and on research into what happens in polluted estuaries such as the Rhine and the Tees, and on how the condition of marine animals can indicate water quality. A research ship, the RRS *Challenger* spent more than a year collecting the physical and chemical data for the model which was planned to become the proving ground for more ambitious projects.

Management of the British Seas

The chapters in Part II of this book have been summarized in Table 14.2. The table again has three columns, but here they are divided into the controls on the distribution of each resource in the region, and into an estimate of the resource models which are presently utilized to develop the resource.

In general, work has only recently commenced on the development of such numerical models and little use is presently being made of the results. For example STWG (1987) describes the development of models over the last 10 years which predict the dispersion of pollutants within the North Sea. The models are presently, however, only capable of large-scale and time-averaged predictions as is illustrated by Plate XIV. The diagram shows one such example of the distribution in concentration (in arbitrary units) resulting from ten continuous sources of passive dissolved tracer. The model incorporates tidal forcing with a constant wind stress and a prescribed diffusion process. The various inputs are weighted as follows: Rhine/Meuse 52 per cent, Elbe 13 per cent, Firth of Forth 6 per cent, Tyne 6 per cent, Weser 5 per cent, Scheldt 5 per cent, Thames 4 per cent, Seine 4 per cent, Humber 3 per cent and Ems 2 per cent. However, since even these relative input figures are not well-established, it is difficult to apply the results to the management of the pollution problem at the present time.

The future undoubtedly lies in the development and utilization of advanced numerical models of the region within an integrated management structure. These models will, on the one hand, provide the information bases for decisions about aggregate extraction, fishing, power generation, and waste disposal problems, and on the other they will provide the opportunities for 'what if' experiments to assess the effects of changing environmental conditions and exploitation patterns. It will only be through the development of such models, sometime early next century, that we shall be in a position to claim that we either understand the oceanography or can manage the resources of the British Seas.

References

Benwell, G.R.R., A.J. Gadd, J.F. Keers, M.S. Timpson and P.W. White, 1971. *The Bushby-Timpson 10 level model on a fine mesh*. Scientific Paper, Meteorol. Office, London, 32, 1–23.

Davies, A.M., 1983. Comparison of computed and observed residual currents during JONSWAP'76. In: Johns, B. (Ed) *Physical Oceanography of Coastal and Shelf Seas*. Elsevier, Amsterdam, 357–386.

Heaps, N.S., 1973. Three-dimensional numerical model for the Irish Sea. *Geophys. J. Roy. Astron. Soc.*, 35, 99–120.

Heaps, N.S. and J.E. Jones, 1983. Development of a three-layered spectral model for the motion of a stratified sea. II. Experiments with a rectangular basin representing the Celtic Sea. In: Johns, B. (Ed.) *Physical Oceanography of Coastal and Shelf Seas*. Elsevier, Amsterdam, 401–465.

Henderson-Sellers, A. and K. McGuffie, 1987. *A Climate Modelling Primer*. Wiley, Chichester, 217 pp.

Huthnance, J.M., 1982. On one mechanism forming linear sand banks. *Est. Coastal Shelf Sci.*, 14, 79–99.

May, R.M. (Ed.), 1984. *Exploitation of Marine Communities*. Springer-Verlag, Berlin, 367 pp.

Pantin, H.M. and C.D.R. Evans, 1984. The Quaternary history of the central and southwestern Celtic Sea. *Mar. Geol.*, 57, 259–293.

Pingree, R.D. and D.K. Griffiths, 1979. Sand transport paths around the British Isles resulting from M_2 and M_4 tidal interactions. *J. Mar. Biol. Assoc. UK*, 59, 497–513.

Prandle, D., 1981. Tidal power schemes in the Bristol Channel and Bay of Fundy. *Wave and Tidal Energy*. BHRA, Bedford, England.

STWG, 1987. Report of the Oceanography Sub-Group of the Scientific and Technical Working Group, 2nd Int. Conf. on the Protection of the North Sea. HMSO, London, 68 pp.

Ziegler, P.A., 1982. Evolution of sedimentary basins in north-West Europe. In: Illing, L.V. and Hobson, G.D. (Eds) *The Petroleum Geology of the Continental Shelf of North-West Europe*. Academic Press, London, pp. 3–39.

Appendix I: Examination questions

This book was written to support an undergraduate course which was assessed through project work and by examination. The following examples are reproduced with the permission of the University of London.

1. Project work

1.1   Draw and describe six cross-sections and a longitudinal section of the Norwegian Channel from the chart and comment on the origins of the feature.

1.2   Prepare a six page application for a licence to prospect for marine aggregate at a site on the UK Shelf, including details of the quality and quantity of reserves and of the economic feasibility of the site.

1.3   Use the wave data at the Varne (Chapter 4) to estimate the likely percentage downtime of a proposed new Channel crossing which is inoperative if H_{max} exceeds 3m.

1.4   Use the peak tidal flow chart (Chapter 5) and the bathymetric chart (Chapter 2) to calculate values of the mixing parameter, M_p (Chapter 6) in the North Sea, and hence draw the line $M_p = 2$ between Flamborough Head and the continent.

2. Examination questions

2.1   Define and explain *either* 100 year significant wave height *or* bedload parting.

2.2   Describe the measurement of tidal elevations and explain the position and progress of the NW Atlantic high water around the UK during two tidal cycles.

2.3 'The coastlines of the British Seas owe more to geological design than geomorphic process.' Discuss the validity of this statement.

2.4 Outline the main seabed sand transport paths in the British Seas and discuss their effect on:
  (a) submarine cable routes
  (b) hydrocarbon pipelines

2.5 Examine the relationship between coastal and offshore fishing areas and the salinity and temperature distributions in the British Seas.

2.6 Compare the economic viability of marine aggregate extraction from the Bristol Channel and the southern North Sea.

2.7 Define and explain *either* dredging prospect *or* salinity.

2.8 Describe (a) the boundaries and shape of the British Seas *and*
  (b) the main bathymetric features of *either* the North Sea *or* the English Channel.

2.9 Explain the evolution of the UK Shelf from the Devonian to the present day.

2.10 Show how the seabed sand transport paths in the British Seas relate to the tidal current streams.

2.11 Describe the passage of the tidal wave around the British Seas, and show how the tidal regime determines the potential distribution of tidal power stations.

2.12 Compare the results of *either* tidal stream *or* wave-height measurements in the western Celtic Sea with those in the Strait of Dover.

2.13 'The distribution of oil tankers in the British Seas is controlled by the time of high water at Dover'. Discuss.

2.14 Explain how the development of more advanced numerical models might assist in controlling the pollution problem in the southern North Sea.

2.15   Compare the wave regime off the west coast of Brittany with that in the eastern Channel.

2.16   Show how the pattern of marine aggregate extraction is controlled by glacial limits in the British Seas.

2.17   Discuss the distribution of one demersal fish species in the North Sea.

2.18   Describe the principal types of oil and gas platform, and explain the distribution of gas fields in UK waters.

Appendix II: Glossary of place names

Compiled, in part, from the British Geological Survey report on the 1:1,000,000 superficial sediment charts.

Anglesey. Island off the north-west coast of Wales. 50°18'N, 04°20'W.

Antrim. County in Northern Ireland. 54°43'N, 06°12'W.

Ardnamurchan. Peninsula off the west coast of Scotland. 56°44'N, 06°10'W.

Aust Cliff. Cliff of Triassic rocks on the south side of the Severn Estuary. 50°34'N, 02°38'W.

Bahama Bank. Sandbank in the Irish Sea, 10km north-east of Point of Ayre. 54°23'N, 04°15'W.

Ballacash Bank. Sandbank in the Irish Sea, 1.5km north-east of Point of Ayre. 54°25'N, 04°14'W.

Barnstaple Bay. North- and west-facing bay on the south side of the Bristol Channel. 51°05'N, 04°20'W.

Barra. Island in the Outer Hebrides. 57°00'N, 07°30'W.

Beaufort's Dyke. An elongated, multiple depression in the North Channel. 54°45'N, 05°15'W.

Bill of Portland. Promontory in the central English Channel. 50°32'N, 02°26'W.

Blackstones Bank. Bank(s) 60km north-west of Islay, corresponding to resistant bedrock. 56°05'N, 07°10'W.

Botney Cut. Elongate depression in the North Sea, 200km east of Flamborough Head. 53°57'N, 03°02'E.

Bristol Channel. East-west stretch of sea between Wales and Devon. 51°20'N, 04°30'W.

Caernarvon Bay. Bay off the north-west coast of Wales. 53°05'N, 04°30'W.

Canna. Island off the west coast of Scotland. 57°03'N, 06°35'W.

Cape Wrath. Most north-westerly headland on the north-west coast of Scotland. 58°38'N, 05°00'W.

Cardigan Bay. Bay off the west coast of Wales. 52°30'N, 04°30'W.

Celtic Deep. In the northern Celtic Sea; a large, shallow depression. 51°20'N, 06°15'W.

Celtic Sea. Sea area between the Republic of Ireland, the coast of

Cornwall, and the 200m isobath. 49°30'N, 07°30'W.

Channel (English). Area of sea between England and France. 50°10'N, 01°30'W.

Coll. Island off the west coast of Scotland. 56°38'N, 06°35'W.

Constable Bank. Sandbank near the north coast of Wales. 53°22'N, 03°45'W.

Cornubia. The peninsula which comprises Devon and Cornwall. 50°30'N, 06°30'W.

Cotentin Peninsula. French peninsula on the south coast of the English Channel. 49°30'N, 01°30'W.

Culver Sand. Tidal sand ridge in the eastern part of the Bristol Channel. 51°15'N, 03°05'W.

Dogger Bank. Large isolated positive feature in the North Sea (summit <20m). 54°45'N, 02°00'E.

East Bank. Area of tidal sand ridges in the North Sea, north-west of Dogger Bank. 55°15'N, 01°20'E.

Edrachillis Bay. Bay on the north-east coast of Scotland. 58°20'N, 05°20'W.

Enard Bay. Bay on the north-west coast of Scotland. 58°08'N, 05°20'W.

English Channel. Sea between England and France. 50°10'N, 01°30'W.

Fair Isle. Island between the Orkney Islands and the Shetland Islands. 59°32'N, 01°37'W.

Fair Isle Channel. The sea between the Orkney Islands and the Shetland Islands. 59°30'N, 01°40'W.

Farn Deeps. Depression in the North Sea, 45km from the north-east of England. 55°30'N, 01°00'W.

Fetlar. One of the Shetland Islands (in the north-east of the group). 60°37'N, 00°52'W.

Fetlar Basin. Structural basin about 15km south of Fetlar. 60°30'N, 00°50'W.

Firth of Clyde. Estuary in south-west Scotland. 55°20'N, 05°00'W.

Firth of Forth. Estuary in south-east Scotland. 56°00'N, 03°00'W.

Firth of Tay. Estuary in south-east Scotland. 56°25'N, 03°00'W.

Fladen Grund. Area in the North Sea, north-east of Aberdeen. 58°32'N, 00°20'E.

Flamborough Head. Chalk headland on the north-east coast of England. 54°07'N, 00°05'W.

Flat Holm. Island in the eastern Bristol Channel. 51°22'N, 03°07'W.

Flemish Bight. (see Southern Bight). General name for the southern part of the North Sea, near Belgium. 51°30'N, 02°30'E.

Forth (River). River in central Scotland. 56°08'N, 04°00'W.

Guernsey. One of the Channel Islands. 49°26'N, 02°35'W.

Haig Fras. Seabed elevation in the Celtic Sea, corresponding to a granite outcrop. 50°13'N, 07°54'W.

Hebridean Shelf. Shelf west of the Outer Hebrides. 58°00'N, 08°00'W.

Humber (River). River in eastern England. 53°38'N, 00°10'W.

Hurd Deep. Largest linear deep in the English Channel; runs about 30km north-west of Guernsey. 49°40'N, 03°00'W.

Inner Hebrides. The islands immediately off the west coast of Scotland, including Skye, Mull, and Islay.

Inner Sound. Stretch of sea running north-south, between mainland
Scotland and the Isle of Skye. 57°25'N, 05°55'W.

Inverness. Town in northern Scotland, at the head of the Moray Firth.
57°28'N, 04°13'W.

Irish Sea (southern). Sea between Wales and Ireland. 52°30'N, 05°10'W.

Irish Sea (northern). Sea between north-west England and Ireland.
53°55'N, 04°20'W.

Islay. Island off south-west Scotland. 55°45'N, 06°10'W.

Isle of Man. Island in the northern Irish Sea. 54°15'N, 04°30'W.

Isle of Wight. Island off southern England. 50°40'N, 01°15'W.

Isles of Scilly. Island group 45km west-south-west of Land's End. 49°55'N,
06°20'W.

Jersey. One of the Channel Islands. 49°13'N, 02°09'W.

King William Banks. Sandbank in the Irish Sea, 25km north-east of Point
of Ayre. 54°26'N, 04°05'W.

La Chapelle 'Bank'. Shelf-edge platform in the southern Celtic Sea
(French sector). 47°37'N, 07°00'W.

Land's End. Furthermost south-east point of the mainland of England.
50°03'N, 05°42'W.

Lewis. Northern island of the Outer Hebrides. 58°10'N, 06°35'W.

Lindis (see Witham). Early name for the River Witham.

Little Minch. Sea between the Isle of Skye and the Outer Hebrides.
57°35'N, 06°55'W.

Lizard Head. Most southerly point of the mainland of England. 49°57'N,
05°12'W.

Lleyn Peninsula. Peninsula in north-west Wales. 52°50'N, 04°40'W.

Loch Fyne. Tributary loch on the north-west side of the Firth of Clyde.
56°05'N, 05°15'W.

Loch Ryan. North-facing bay in the Firth of Clyde. 54°56'N, 05°00'W.

Lobourg Channel. Linear deep on the French side of the Strait of Dover.
50°55'N, 01°30'E.

Luce Bay. South-east-facing bay in the northern Irish Sea. 54°45'N,
04°50'W.

Lune Deep. Narrow submarine channel in the entrance to Morecambe
Bay. 53°55'N, 03°10'W.

Lyme Bay. South-facing on the north side of the English Channel.
50°35'N, 03°00'W.

Mainland (1). Largest of the Orkney Islands. 59°00'N, 03°15'W.

Mainland (2). Largest of the Shetland Islands. 60°15'N, 01°30'W.

Malin Head. Headland on the north coast of Northern Ireland. 55°22'N,
07°24'W.

Malin Sea. Sea between the north coast of Ireland and the western islands
of Scotland. 55°50'N, 07°30'W.

Markhams Hole. Depression in the North Sea, 150km north-east of Great
Yarmouth. 53°50'N, 02°40'E.

Middle Bank. Bank between Islay and the northern coast of Ireland.
55°28'N, 06°24'N.

Minch. North Minch plus Little Minch.

Moray Firth. North-east-facing bay in north-east Scotland. 58°00'N, 03°00'W.

Morecambe Bay. Bay on the west coast of north-west England. 54°00'N, 03°00'W.

Muck. Island off the west coast of Scotland. 56°50'N, 06°15'W.

Muck Deep. Depression 15km south-west of Muck. 56°48'N, 06°34'W.

Muddy Hollow. Shallow, elongated depression, south of the Lleyn Peninsula. 52°45'N, 04°27'W.

Mull. Island off western Scotland. 56°25'N, 06°00'W.

Norfolk Banks. North-west to south-east trending tidal sand ridges in the southern North Sea. 53°10'N, 02°20'E.

Normandy. Province in northern France. 49°00'N, 00°00'E.

North Channel. Sea between Scotland and Northern Ireland. 54°50'N, 05°30'W.

North Minch. Sea between north-west Scotland and the Outer Hebrides. 58°00'N, 06°00'W.

North Rona. Islet about 75km north-west of Cape Wrath. 59°08'N, 05°50'W.

North Sea. Sea between eastern Britain and the north European mainland. 56°00'N, 04°00'E.

Northern Ireland. Part of the United Kingdom lying in the north-east part of Ireland. 54°30'N, 06°00'W.

North Uist. One of the Outer Hebrides. 57°35'N, 07°20'W.

Northern Shelf. Shelf to the north-west and north of Sutherland.

Norwegian Trench. North–south orientated deep, separating the main North Sea shelf from the Norwegian shelf. 60°00'N, 04°00'E.

Orkney (Islands). Island group, centred some 40km north of the Scottish mainland. 59°00'N, 03°00'W.

Outer Hebrides. Elongated island group, centred some 60–90km west of the Scottish mainland. 58°00'N, 07°00'W.

Outer Silver Pit. Depression in the North Sea, 160km north-east of Great Yarmouth. 54°00'N, 02°25'W.

Pentland Firth. Strait between the Orkney Islands and the Scottish mainland. 58°45'N, 03°10'W.

Pobie Bank. Bank 52km east of the Shetland Islands. 60°30'N, 00°05'W.

Point of Ayre. Most northerly point of the Isle of Man. 54°24'N, 04°23'W.

Raasay. Island between Skye and the Scottish mainland. 57°25'N, 06°03'W.

Rathlin Island. Island 5km off the north coast of Ireland. 55°16'N, 06°10'W.

Rhine (River). Major river, flowing into the Flemish Bight from western Europe. 52°00'N, 06°00'E.

Rhum. Island off western Scotland. 57°00'N, 06°20'W.

Rubha Coigeach. Headland in north-west Scotland. 58°06'N, 05°25'W.

Sandettie Bank. Tidal sand ridge in the Flemish Bight. 51°10'N, 01°50'E.

St Bees Head. Headland on the north-west coast of England. 54°30'N, 03°37'W.

St Catherine's Deep. Linear deep, south of the Isle of Wight. 50°33'N,

01°13'W.

St George's Channel. Western part of the southern Irish Sea. 52°20'N, 05°40'W.

St Kilda. Island west of the Outer Hebrides. 57°48'N, 08°34'W.

St Magnus Bay. Bay on the west coast of the Shetland Islands. 60°23'N, 01°35'W.

Sand Bay. Bay in the eastern Bristol Channel. 51°20'N, 03°00'W.

Sand Hills. Tidal sand ridges north-east of Flamborough Head. 54°15'N, 00°30'E.

Sea of the Hebrides. Area south-west of the Isle of Skye, between the Outer Hebrides and the Scottish mainland. 56°50'N, 07°10'W.

Seven Stones. Reef (granite) off the south-west coast of Cornubia. 50°02'N, 06°07'W.

Severn Estuary. Estuary of the River Severn. 51°33'N, 02°45'W.

Severn (River). River in south-west flowing into the Bristol Channel. 51°30'N, 02°50'W.

Shetland (Islands). Island group, centred some 230km north-north-east of the Scottish mainland. 60°20'N, 0130W.

Skerryvore. Skerries on crest of south-east-facing scarp, south of Tiree and Coll. 56°18'N, 07°08'W.

Skye. Island off the west coast of Scotland. 57°14'N, 06°00'W.

Smiths Knoll. Tidal sand ridge in the Flemish Bight. 52°52'N, 02°15'E.

Sound of Raasay. Large inlet east of Skye. 57°25'N, 06°07'W.

South Falls. Tidal sand ridge in the Flemish Bight. 51°20'N, 01°48'E.

South Harris. Peninsula in the Outer Hebrides, north-west Scotland. 57°50'N, 07°00'W.

South Rona. Island between the Isle of Skye and the Scottish mainland. 57°33'N, 06°00'W.

South Uist. One of the Outer Hebrides. 57°15'N, 07°20'W.

Southern Bight. Synonym for Flemish Bight.

Southern Trench. East–west-trending deep on the southern side of the outer Moray Firth. 57°45'N, 02°08'W.

Spurn Head. Spit of sand and gravel at the mouth of the River Humber (north side). 53°36'N, 00°08'E.

Stack Skerry. Islet 50km north-east of Cape Wrath. 59°02'N, 04°30'W.

Stanton Banks. Highly irregular banks north-west of Islay, corresponding to resistant rock masses. 56°20'N, 08°30'W.

Steep Holm. Island in the eastern Bristol Channel. 51°20'N, 03°07'W.

Strait of Dover. Strait separating the English Channel from the Flemish Bight. 51°00'N, 01°30'E.

Sula Sgeir. Islet and reefs about 85km north-west of Cape Wrath. 59°05'W, 06°10'W.

Sule Skerry. Islet about 60km north-east coast of Cape Wrath. 59°05'N, 04°24'W.

Swallow Pit. Depression in the North Sea, 110km from the coast of north-east England. 56°00'N, 00°00'E.

Tampen. Bank 200km north-north-east of Shetland. 60°50'N, 01°15'E.

Tampen Ridge. Asymmetric ridge running south-south-east of Tampen.

Tay (River). River in eastern Scotland, entering the North Sea. 56°26'N, 03°25'W.

Thames (River). River in southern England, entering the Flemish Bight. 51°30'N, 00°00'E.

Thames Estuary. Estuary of the River Thames. 51°30'N, 01°00'E.

Tiree. Island off the west coast of Scotland. 56°30'N, 06°55'W.

Trawling Ground. Elongate zone of muddy sediment in Cardigan Bay. 52°17'N, 04°15'W.

Tremadog Bay. South-facing bay at the northern end of Cardigan Bay. 52°50'N, 04°15'W.

Tyne (River). River in north-east England, flowing into the North Sea. 55°00'N, 01°25'W.

Unst. Most northerly of the Shetland Islands. 60°45'N, 00°55'W.

Unst Basin. Elongate structural basin, running north–south about 12km east of the Shetland Islands. 60°35'N, 00°35'W.

Wash, The. North-east-facing bay on the east coast of England. 52°55'N, 00°20'E.

Wear (River). River in north-east England. 54°54'N, 01°25'W.

Witch Ground. Area of muddy sediments in the North Sea, 150km north-east of Aberdeen. 58°00'N, 00°00'E.

Witham (River). River in southern Lincolnshire. 52°58'N, 00°00'E.

Ystwyth (River). River in central Wales. 52°21'N, 04°00'W.

Index